Homeopathy for the Herd

A Farmer's Guide to Low-Cost, Non-Toxic Veterinary Care for Cattle

C. Edgar Sheaffer, V.M.D.

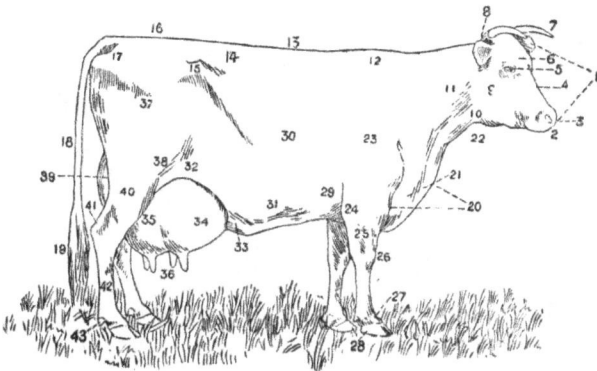

Homeopathy for the Herd

A Farmer's Guide to Low-Cost,
Non-Toxic Veterinary Care for Cattle

C. Edgar Sheaffer, V.M.D.

Acres U.S.A.
Greeley, CO

Homeopathy for the Herd
A Farmer's Guide to Low-Cost,
Non-Toxic Veterinary Care for Cattle

Copyright ©2003 by C. Edgar Sheaffer, V.M.D.

Acres U.S.A.
P.O. Box 1690
Greeley, CO 80632 U.S.A.
800-355-5313 • 970-392-4464
info@acresusa.com • www.acresusa.com

Printed in the United States of America

Pulisher's Cataloging-in-Publication

Sheaffer, C. Edgar, 1943-
Homeopathy for the herd: a farmer's guide to low-cost, non-toxic
   veterinary care for cattle. GREELEY, CO: ACRES USA, 2003

   xiii, 206 pp. charts, tables.
   Includes index

   ISBN: 1-601731043

   1. Homeopathic veterinary medicine.   2. Cattle—Diseases—
Alternative treatment. I. Sheaffer, C. Edgar, 1943-. II. Title.

   SF961.   636.20895532

Contents

Foreword

This book is more than the latest advances in veterinary homeopathic medicine that are a challenge to verify from the perspective of scientific materialism and reductionism. This is a perspective that sees animals as machines and disease as a cause in itself and never as symptom or consequence. This book is more than the latest advances in animal production science, and most animal scientists will see it as heretical and even nonsensical. What kind of production are we talking about and under what economic, social and climatic conditions? This book has little place in conventional intensive cow-calf, dairy and beef operations that rely on tons of costly and potentially harmful pharmaceutical products and agrochemicals. But it should take center-place, since the author, a veterinarian with a strong and cost-saving bias toward preventive medicine, addresses the physical, environmental and behavioral needs of cattle in an ecological context of humane and sustainable production that is profitable. His methods are good for the animals, the land, the producer and the consumer.

Veterinary homeopathy, like other alternative and supplemental therapies and preventive medicine/health maintenance practices, works best in the kind of holistic framework that Dr. Sheaffer provides. This framework includes the basic bioethical principles of animal husbandry, health and well being: right environment; right nutrition; right handling and understanding (respect and compassion); and right breeding. No breed or line of cattle or other livestock is ideal for all conditions. Hence the need to preserve rare breeds that are now endangered by the predominance of a few breeds and lines developed for intensive, bio-industrial systems of dairy and beef production. Species diversity as well as genetic diversity makes for healthier ecosystems. This is another holistic veterinary and animal husbandry principle that recognizes increased efficiency of land and forage use and disease control by mixed and rotational grazing practices that maximize each species' biological value and contribution to the farm or ranch polyculture, including cattle, sheep goats, pasture and wood-lot pigs, and free-range tick and fly controlling chickens.

The effectiveness of various homeopathic preparations in the prevention and treatment of various diseased conditions in cattle is supported by clinical evidence and hands-on, in-field experiences. Those people who want to explore bovine homeopathy as an alternative or adjunct to conventional (allopathic) veterinary practices will find this book an invaluable guide and resource. But as the author advises, use the homeopathic approach first and not as a last result for optimal results.

We live in times of increasing uncertainty locally and globally. Such times call for a combination of caution, courage, commitment and creativity. This book combines those virtues with the traditional values of compassionate and responsible land stewardship and animal care. In so doing, it connects us with a deeper wisdom, now almost lost, that equates healthcare holistically with earth care and animal care.

As history informs, the extensive (free-range) production of cattle and other domestic farmed animals has too often neglected earth care, with consequent ecosystem degradation and collapse through overgrazing, overstocking and other poor husbandry practices that also resulted in lowered productivity and livestock diseases, especially parasitic and infectious, often related to chronic malnutrition and immune system impairment. The spread of diseases from livestock to wildlife, coupled with relentless and often indiscriminate predator 'control' (i. e., extermination) have resulted in irreparable loss of biodiveristy and natural habitat in many parts of the world.

The more recent adoption of intensive, industrial-scale dairy and beef feedlot and veal confinement operations is supported by a nexus of often far away cow-calf producers, and by vast acres of corn and soy monocultures and other livestock feedstuffs, including the byproducts of the human food and beverage industries. A full cost-accounting of conventional beef and dairy industries heavily subsidized by the government, to the benefit of primarily of the 'agribusiness' livestock feed, drug and petrochemical multi-nationals, is long overdue. Its complexity conceals many indirect costs, notably public health (especially from harmful bacteria and over-consumption of animal fat and protein); animal health and welfare (especially from stress-related 'production' diseases); adverse environmental impact from loss of biodiversity and to pollution, especially of ground and surface waters; and adverse impact on rural communities and the once vibrant nexus of diverse family farms and ranches. The former have been displaced by the economies of scale, the politics of agri-greed and public indifference; and the scientific myopia and biological illiteracy of academia caused in part by production-oriented agronomics, animal science and agri-biotechnology that corporate interests and a competitive, rather than cooperative, world market have nurtured into a cancer of the soul of agriculture.

America, in its renewed quest for national security, will continue to chase phantoms until food and water quality and security, and energy sustainability and efficiency are higher on the nation's agenda—which means restoring the heartland, rural communities, bioregional economies and the family farm.

We do not need a cattle industry that condones the confinement and slaughter of the last of the great buffalo herds, and that transports, at great cost and animal suffering, cattle of different ages to and from feedlots to ultimate slaughter, spreading diseases in the process. With its over-centralization of processing it has become a threat to public health and is highly vulnerable to terrorist attack with potentially devastating economic and social consequences, that could parallel those of the United Kingdom's mad cow and recent foot and mouth disease outbreaks.

Dr. Sheaffer's book should be a standard text for critical discussion in all relevant state university departments of agriculture, animal science and, of course, veterinary medicine. Non-governmental organizations like Heifer Project International, and government organizations such as the FAO (Food and Agriculture Organization of the United Nations), should conduct in-field evaluations of bovine homeopathy and indigenous, traditional medicines that may prove far more beneficial and cost-saving than exclusive reliance on allopathic medicines, like antibiotics, steroids and genetically-engineered hormones and vaccines.

This book's beauty and ultimate purpose is not to sell homeopathy as a veterinary panacea. Rather, the author puts veterinary homeopathy in perspective, as it should be, of whole-herd healthcare maintenance and disease prevention, especially parasitic and various contagious diseases and infections that involve organisms that are becoming resistant to conventional vermicides, insecticides and bactericidal drugs. Since homeopathic medicines are all natural, and since in extreme dilution they can be of no risk to consumers, then there can be no scientifically valid grounds for them not to be acceptable under the National Organics Standards. Finally, this book will be enjoyed by those farmers, ranchers and their veterinarians who have, or are seeking to 'go organic', or to adopt other ecologically sustainable, economically viable, humane and socially just production systems.

— *Michael W. Fox, B. Vet. Med., Ph.D., D.Sc., M.R.C.V.S.*

Preface

I graduated from Gettysburg College in 1965, did graduate work in nutrition at Penn State and then graduated from the University of Pennsylvania School of Veterinary Medicine in 1970. I began practice as a traditional veterinarian at that time. After ten years of practice, I felt that there had to be a better way to practice medicine than the way I was doing it. I began by using fewer drugs and more vitamins and studying nutrition. Then a flyer appeared on my desk announcing that George Macleod, M.R.C.V.S., from England, would be lecturing on veterinary homeopathy at nearby Millersville University. My employer at the time had no interest in going, but permitted me to attend. That meeting changed my life and started me on the journey toward using homeopathic medicines in my practice.

These days I use as much homeopathic medicine as possible. Actually, about 95 percent of my practice is homeopathic, and the remaining five percent is conventional. I always try to think of a homeopathic medicine before I use a stronger, conventional method of therapy. On rare occasions when I can't get results with a homeopathic medicine, antibiotics are helpful. Where there is obvious bacterial infection and sensitivity studies indicate a particular antibiotic, I prescribe it along with the indicated homeopathic medicine. Guy D. Hoagland, M.D. of Melbourne, Florida, states it well, "Homeopathy as a science can build on the conventional practice of medicine and provide more than we can provide without homeopathic medicine."

Homeopathy is a part of the whole, not the whole of medicine in itself. In medicine, one needs to be aware of nutrition, environmental stresses, vaccination history and genetic predisposition, to name a few factors. Most homeopathic medicine is given orally as opposed to giving an injection. The medicine is given right in the mouth, right on the tongue or in the cheek. That is the ideal method. When that is not possible, the tablets can be placed directly on top of a small amount of grain. The medicines are given on the basis of a total symptom picture. It takes a while to learn the picture of each medicine, and there are between two and three thousand homeopathic medicines in common use in the world, so it can seem overwhelming at first. There is some overlap, but there are also distinctions, and the important thing is to learn those distinctions. A person learning homeopathy for the first time studies a few medicines at a time. As he or she becomes more familiar with those, confidence increases. Using those few medicines leads to learning about other related medicines. It's learning through experience.

With conventional medicine, you have basically three main options: antibiotics, anti-inflammatories (steroids, etc.), and surgery. With homeopathy and nutritional therapy, the options are almost unlimited. My mentor, Dr. Macleod, said about homeopathy, "It always works." For homeopathic medicines to be effective, the patient must be impacted with the energy of the medicine. There may be factors which prevent that, such as persistent viremia (viruses in the blood), as found in BVD and parvo virus, and abnormal endocrine functions. Persistent viremia often responds to the appropriate homeopathic nosode (a remedy produced from a product of the disease) given in combination with the simillimum (the homeopathic remedy which most exactly reproduces the symptoms of any disease). Endocrine dysfunction may be treated using a combination of homeopathy, nutritional therapy and a natural supplement. A natural supplement may target any specific organ or tissue of the body. Thyroid and adrenal supplements are often prescribed for animals.

This book is intended to help you get started in the use of homeopathic medicines with cows. If you get a homeopathic kit and start using the medicines for first-aid needs a little at a time, you will see results. I hope that you will try some of the medicines found in this book and keep a journal of those used and the results with your animals. I assure you that you will learn something new from each patient and case.

My family has always supported my sojourns in veterinary medicine. Both Amy and Clark would ride along on calls and soon came to the decision, "Dad, that's just too much blood." Rather than veterinary medicine, both chose the educational field as their vocation. They were the "guinea pigs" when I was learning homeopathy. Everything from bruises and abrasions, colds and influenza, insect stings, ingrown toenails, sore throats and sports injuries to anxiety before exams and indigestion from college foods were treated with homeopathic medicines. Our children learned the value of these medicines first hand and now reach for the homepathic medicines first. When the youngest granddaughter has an abrasion, she asks, "Mommy can I have a 'bam-baid' and CalenDUla?"

1.

Introduction to Homeopathy

Farming has been in my family for generations. My first job in the barn at the age of four was to milk one Guernsey cow. My father and mother and their parents before them milked cows and plowed the rich limestone land of southeastern Pennsylvania. My family's farm provided a wonderful place for my brother and I to grow and learn the standards of ethical and quality living. I felt that I could best support that way of life by choosing a career in veterinary medicine.

"Physician do no harm." This was the admonishment of Hippocrates, father of modern medicine. Dr. Christian Frederick Samuel Hahnemann, discoverer of homeopathic medicine, said that the "highest goal and the only goal of the physician is to cure, to heal the sick." Homeopathic medicine is a powerful tool that we have at our disposal which can heal without doing harm. For the farmer and the veterinarian, this system offers a gentle and natural method of healing.

Two of the goals of my veterinary practice are to offer gentle, natural and non-invasive therapies, and to provide veterinary care at an affordable cost. The best way to do this is to use natural, non-chemical products, specifically homeopathic medicines. Homeo-pathic medicines are safe—safe for the environment, safe for the farm family, safe for the consumer and safe for the farm financial picture.

Homeopathic medicines are especially suited for herds of animals. A practitioner may either choose to prescribe for the individual cow or to view the herd as a whole. Either choice produces beneficial results. Homeopathic medicine provides both environmental and farm safety, produces no adverse side effects, causes no immune suppression, increases the vitality of the succeeding generations, provides an option for the treatment of chronic disease, is cost effective and, most importantly, produces no drug residues. A farmer can sleep peacefully knowing that the milk in his/her bulk tank contains no forbidden drugs. Once the practice of first using natural medicines is adopted, there is rarely a need to administer drugs which require the withdrawal of milk, meat and eggs from the marketplace.

Research in veterinary homeopathy is limited, but clinical anecdotal evidence exists to indicate that veterinary homeopathy may be beneficial. It is recommended that further research be conducted in veterinary homeopathy to evaluate efficacy, indications, and limitations. Since some of these substances may be toxic when used at inappropriate doses, it is imperative that veterinary homeopathy be practiced only by licensed veterinarians who have been educated in veterinary homeopathy.

Two Principles of Homeopathy

1. Any medicine strong enough to cure is also strong enough to cause harm.
2. Any medicinal substance may either suppress or palliate or cure.

Palliation vs. Cure

When a patient is recovering, symptoms will move from the more vital organs to the less vital ones and from the center of the body outward. Vital organs are heart, brain, liver, lungs and kidneys. Less vital are sinuses, skin, hair and extremities.

When livestock being treated homeopathically are in a recovery process, the center of the disease will move toward the extremities. It is very common for a horse recovering from chronic obstructive pulmonary disease (COPD or heaves) to become lame in one front limb due to an inflamed joint or hoof. Likewise, it is common for a cow recovering from pleuritis to develop a sore hoof or an inflamed quarter. If one is not aware of the healing process, suppressive therapy may be instituted which will cause the original disease to reappear (antibiotics, steroids and other anti-inflammatories may do this).

Several definitions will be helpful at this point. Palliation is the act of relieving suffering without actually curing the patient. The palliated patient will usually take some medication daily over several months. An example is a patient with thyroid disease needing supplemental therapy. The thyroid medication will not heal the patient, but suffering is relieved with appropriate daily dosing to support thyroid function. Thus palliation supports the patient in his disease.

Suppression is the opposite of healing. The disease moves from a less vital organ to a more vital organ. Until the process is reversed, the patient can never become well. A common example of suppression in dogs occurs when a cortisone salve is applied to a skin rash. The skin improves but within weeks a nearby joint becomes inflamed as the result of the topical steroid therapy. The disease has been suppressed into a deeper organ. Additional examples of suppression have been observed after DMSO administered by any route, topical steroids in the ears, and antibiotics in the mammary gland.

Palliation may be defined as a temporary relief from suffering. Cure indicates that there is a permanent removal of disease. Our goal should always be to cure. Only when cure is deemed impossible should we use homeopathic medications for palliation. Because the body heals from the more vital to the less vital organs and tissues we should never suppress. A suppressed disease will disappear for a short time, only to be manifested in a more life threatening condition later.

A cow is treated with mastitis tubes at drying off. She may freshen without any apparent problem only to develop kidney or liver disease two to four months later. Could it be that mastitis tubes are suppressive? After millions or perhaps billions of intra-mammary infusions, we are no closer to curing mastitis by this method. As a wise little boy once asked, "Hey mister, don't you see that there's a cow attached to that udder?" Treat the whole patient and your success rate will increase. Treat only the part and failure will surely result.

What is a Homeopathic Medicine?

Homeopathic medicine developed from one doctor's frustration and curiosity. Christian Frederick Samuel Hahnemann became frustrated with the practice of medicine in the 1700s. Treatments were harsh, often leaving the patient in worse shape after treatment than before the drugging. For example, one popular treatment for syphilis was to apply elemental mercury salve to one quarter of the body once weekly for four weeks. Applying the salve to more surface area than that would likely kill the patient from mercury poisoning. After several of these applications the patient's gums and tongue would swell. The next stage in this toxic therapy resulted in the teeth falling out. When the physician observed the severe gingivitis with loose and lost teeth, he would pronounce the patient "mercerized" and cured.

Hahnemann was revolted by such treatments and actually closed his medical practice. But he did not give up on finding ways to help his fellow man. He believed that God in His beneficence must have revealed to someone a gentle and effective method of treating the sick. Even though he himself was no longer in clinical practice, Hahnemann tirelessly continued the search. Samuel was married and had a large family. To provide for his wife and children, he took up the translation of medical textbooks. It was there that he began to find the answer to his deeply held curiosity. While translating Cullen's *Materia Medica* from English into German, he became interested in the relationship between the curative action of Cinchona (bark) and its toxic properties. His enquiries led to an experiment which he performed on himself.

"I took for several days, as an experiment, four drams of good China [*Cinchona officinalis*] twice daily. My feet and fingertips, etc., at first became cold. I became languid and drowsy; then my heart began to palpitate; my pulse became hard and quick; an intolerable anxiety and trembling (but without rigor); prostration in all the limbs; then pulsations in the head, redness in the cheeks, thirst . . . all of those symptoms which are typical of intermittent fever . . . .This paroxysm lasted for two or three hours every time, and recurred when I repeated the dose, and not otherwise. I discontinued the medicine and I was once more in good health."

This was the first recorded "proving" of a substance in the medical literature. Hahnemann intentionally gave the poisonous substance to himself to find what the medicinal properties of Cinchona bark might be. When he took the China he developed symptoms, and when he did not repeat it the symptoms went away. Being an excellent scientist, Hahnemann quickly perceived the cause and effect. A lesser man would have believed that he was coming down from an illness and aborted the experiment. Proverbs 4:7 says, "In all thy getting, get wisdom, and with wisdom, understanding." Hahnemann was wise enough to know that the symptoms he experienced were not originating from his body, but from the Cinchona bark itself.

Soon he was testing other natural substances from the *Materia Medica* of other medical traditions that were reported to possess healing properties. He recorded all symptoms that he experienced while taking each substance. Confirmatory provings were completed by his family, friends, and professional colleagues. He understood that this discovery would be controversial because it ran contrary to the allopathic medical teaching of his time. But he believed a greater purpose was involved. Hahnemann's deep faith in the Creator, like that of his fellow countryman Martin Luther before him, drove him toward this new discovery. The searching of both famous Germans revealed eternal truths that had previously not been widely considered. For Luther and for Hahnemann, God's purpose always had ultimate preeminence.

In 1828 Hahnemann wrote, "If I did not know for what purpose I was put here on the earth—to become better myself as far as possible and to make better everything around me, that is within my power to improve—I should have to consider myself as lacking very much in worldly prudence to make known for the common good, even before my death, an art which I alone possess, and which it is within my power to make as much profit as possible by simply keeping it a secret." [*The Chronic Diseases, Their Peculiar Nature, and Their Homeopathic Cure,* Boericke and Tafel, Philadelphia, PA, 1896.]

Thirty-eight years after Hahnemann's discovery, many of his professional colleagues still refused to accept the findings as valid. He found that natural medicines act dynamically rather than materially. Symptoms produced by material doses of a medicinal substance can be removed by the same substance if given homeopathically. The correct prescription is found when the symptom picture of the patient matches the symptom picture of a medicine, and each homeopathic medication was shown to possess a specific symptom picture.

Two hundred and twelve years later, there remain two points of view concerning Hahnemann's discovery. The homeopath works from the point of view that "like curses like." This point of view is referred to as the *simillimum*. The allopath's approach is to come against the disease by prescribing anti-pyretics, anti-inflammatories, and antibiotics in material doses. Occasionally a homeopath may prescribe in material doses but, unlike the allopath, he looks for the dynamic or energetic effect.

Veterinary homeopathy is a medical discipline in which conditions are treated by the administration of substances that are capable of producing clinical signs in healthy animals similar to those of the animal to be treated. These substances are used therapeutically in minute doses.

What is a Nosode?

The term nosode may be defined as a product of disease that is obtained from any affected part of the system in a case of illness and thereafter potentised. For example, a Fowl Pest nosode is made from the respiratory secretions of affected birds. In specific cases of bacterial, viral or protozoal disease, the causative organism may or may not be present in the material, and the efficacy of the nosode in no way depends upon the organism being present. The response of the tissue to invasion by bacterial or viruses results in the formation of substances which are in effect the basis of the nosode. In shirt-sleeve English, this means that the mother tincture used to make a nosode contains both the material from the patient and a sample of the energy from the patient responding to the illness, whether or not the organism is present.

All nosodes are prepared in homeopathic pharmacies according to strict pharmaceutical standards, and all nosodes have a "prescription only" label. Fortunately, the number of veterinarians using homeopathic medicines are increasing yearly and as a result clinical data on nosode use is also increasing. Some nosodes have undergone extensive provings as in the case of *Lyssin*, the rabies nosode. Others have not as yet been thoroughly proven, but may be used extensively in clinical situations.

Why use Homeopathy?

When we use homeopathic medicines on the farm, we begin to look at symptoms in a different light; they are no longer something to be feared, suppressed or destroyed. They are pieces to a puzzle. The symptoms of the patient are extremely helpful to the vet or farmer in choosing a homeopathic medication. One could say that illness becomes an opportunity to improve the health of the patient. The proper homeopathic medicine augments the healing response of the patient's own immune mechanism. After the patient recovers from an illness, she is more resistant to the illness than before symptoms first became manifest.

The most important point is to "treat with homeopathy first." An animal weakened by prolonged illness or treated with many chemical drugs has an almost impossible task to return to health. They have been taken down a road of immune suppression and their bodies are filled with toxic substances. To "try" homeopathic medicines as a "last resort" is not the best approach for anyone. The results are usually not satisfactory and everyone experiences frustration, including the "natural vet." All persons involved in the care of animals must be diligent in the use of knowledge and common sense. Nothing is impossible for those willing to venture forth and use what they have learned.

While the correct homeopathic medicine acts to stimulate the patient's immune system, it has been documented that the following classes of medicines can act as immune suppressive agents: prostaglandins, antiprostaglandins, anti-inflammatories, antibiotics, anti-convulsants, anesthetics, steroids, and vaccines. Ideally, agriculture should be carried on without drugs; however, there are times when sick animals require treatments. Homeopathic medicines give the farmer and his/her veterinarian options and a new freedom.

The substances (animal, vegetable, or mineral) used for homeopathic medicines are diluted serially either in 1/10th stages (decimal dilutions) or 1/100th stages (centesimal dilutions) which are designated X and C potencies respectively. Thus *Arnica montana* 12X is tincture of *Arnica* diluted one in ten, twelve times or a dilution of 10^{-12}. Succussion or shaking is carried out at each stage. It is succussion which releases the curative energy of the substance, while the dilution process removes its toxic or harmful effects.

When farm animals are treated solely with natural homeopathic medicines, there is no chemical residue in the meat, eggs or dairy products. For example, beef cattle at the time of castration or dehorning tolerate the procedures better when *Aconitum napellus* (Monkshood) is administered before surgery and *Arnica montana* (Leopard's Bane) is given following the surgery. Administering medications to all animals can be accomplished by dissolving eight to ten pellets in one-half to one ounce of distilled water. The process requires approximately one minute; the resulting solution is administered orally. Some of our farmers prefer using a small pill syringe and dosing the same amount of pellets to each animal. When the animals are in the chute, *Arnica* 12X, 12C, 30X, 30C, 200X, 200C or 1M can be given before the animal is released. *Arnica* appears to facilitate healing, lessens pain and restlessness, decreases blood loss and lessens the effects of shock due to the surgical procedure. *Arnica* is probably the most popular of all homeopathic medicines and is very effective. It leaves no residue in the beef unlike the antibiotics and hormones that are routinely administered to beef cattle.

Perhaps the best way to show how homeopathy works is to give you examples from my own practice. Dolly's story may be extreme, but I have seen just this sort of success many times.

I was first called to attend Dolly on a Friday evening. The heavy-producing dairy animal was hemorrhaging from her left nostril. She was excited and fearful with a necrotic (rotten) odor to her breath. The water bowl and manger were filled with bright red semi-clotted blood. The previous veterinarian had given several injections of vitamin K, which had not resolved the problem. Quite obviously, I was dealing with a life and death situation. I suggested immediate slaughter to relieve suffering and salvage the meat, but was told that it was out of the question. My attempt at packing the left nostril only caused the blood to pour forth from the right nostril. Using an epinephrine nasal spray was also unsuccessful. I then asked the dairyman if he would like to try a homeopathic medicine. He replied, "Let's do it. Nothing else is working."

The prescription of *Phosphorus 200C* was given orally every 15 minutes. Much to our surprise and relief, we observed the gush decrease as the blood began to clot. By the third dose, there was only a small dripping of blood and by the fifth dose, the bleeding had stopped completely. I instructed the dairyman to watch her closely over the weekend and to call if any symptoms reappeared.

Despite the loss of several quarts of blood on Friday, Dolly was back eating full ration on Sunday when her production was only 30 pounds. On Monday the dairyman proclaimed her production to be 60 pounds and she was eating and drinking all that was offered. Dolly stayed in the herd four more years and held the herd production record throughout that time. She gave the farm family four more calves. Her health was excellent and she required no further medication—homeopathic or otherwise.

We have observed this phenomenon of immune stimulation over and over again in many patients. For example, a cow is diagnosed with a displacement (twisted

stomach). The dairyman chooses to treat with homeopathic medicine instead of immediate surgery. The recovery rate of cows treated homeopathically for a displacement is in the range of 50-75 percent. If the cow still
needs surgery two to three days after homeopathic treatment, her recovery is more rapid than those that have been
on steroids or other non-homeopathic drugs.

Crushing Chronic Disease in the Herd

Samuel Hahnemann, M.D., compiled his philosophies, his love
and pursuit of wisdom and search for basic principles and life's
attitude into *The Organon of Medical Art*. Wendy Brewster O'Reilly,
Ph.D. edited and annotated the sixth edition of *The Organon* in
1966, and it is from this work that the following "points" are
presented.

*Point: Allopathic treatments are like acts of war: the land (the organism)
is destroyed but the enemy remains.*

In agriculture, herbicides kill the weaker weeds but the fertility
of the soil is not enhanced. Antibiotics destroy some pathogens
and many beneficial organisms, further weakening the patient
whose liver must now detoxify the antibiotic.

*Point: He is likewise a sustainer of health if he knows the things that
disturb health, that engender and maintain disease, and is aware of how to
remove them from healthy people.*

Homeopathic/holistic veterinarians must learn with every
farm, every case, every situation those things that are disturbing
health and that engender and maintain disease, such as feeding of
abnormal food stuffs (meaning animal parts, candy, cappuccino,
chicken manure), damp barns with poor ventilation, uncontrolled
current shocking the livestock, and inadequate bedding, to name a
few. Feeding roughage instead of grains to beef cattle for one or
more weeks before slaughter virtually eliminates deadly *E. coli*
infection in humans. Yet, conventional wisdom would rather irra-
diate meat than prevent the problem by changing the feeding
practices.

*Point: When all the symptoms of disease have been lifted, the disease is
also cured in the interior.*

The interior disease referred to by Hahnemann is the sub-clinical disease that becomes manifested during stress. For example, a dairy cow accustomed to a TMR ration and confined environment becomes thin and drops milk production when moved to a grazing situation. She is not in physical shape to walk and chew and digest all at the same time.

Point: In health, the life force keeps all parts of the organism in harmony.

The animal's harmony is directly related to healthy balanced soil, the elimination of suppressive drugs, available clean water, abundance of grass, adequate grazing time and sound husbandry practices.

Point: An epidemic of acute disease seizes many people at the same time in a very similar way, typically in places where people are thickly crowded. All cases in a given epidemic have the same origin.

Are we seeing this type of health situation with pigs in commercial farrowing pens not allowed access to the earth, chickens in commercial poultry houses, cows in confinement and possibly children in daycare centers under artificial lights, breathing recirculated air, eating irradiated genetically engineered, highly preserved, processed, microwaved, CLA-deficient foods?

Point: Chronic diseases caused by prolonged, violent allopathic treatment are the most incurable.

Nothing is so frustrating and difficult as a case or a herd that is presented for homeopathic consultation and evaluation that has been given every allopathic drug known to man. Is it any wonder that the caring veterinarian or healthcare provider wants to throw his or her hands up in defeat?

Start with the natural first. One can always proceed to the antibiotic or other "anti" drug or surgical intervention if the medical/natural treatment is of no affect.

Point: Start with small doses so that no harm will be done if the medicine given proves to be an unfitting one. Medicines may be administered by olfaction, inhalation and through the skin.

When giving low potencies, one can determine within a short period of time whether the animal is responding favorably. If not, one can stop the first medicine, review the symptoms again and choose another medicine. The 6C, 12C, 12X, 30C, 30X and occasionally 200C or 200X is selected for use in livestock and food-producing animals. At those energy levels, none of the material product is in the preparation and therefore no drug residues can occur.

Scientific Evidence

Most homeopathic research in livestock has been performed using clinical field trials as opposed to laboratory trials, which explains why so many claim that homeopathy is not 'real' medicine in spite of the anecdotal evidence that supports the claims made by homeopaths. There have been published scientific accounts describing the efficacy of homeopathic methods. Dr. Christopher Day, author of *The Homeopathic Treatment of Beef and Dairy Cattle,* studied bovine mastitis using a combined mastitis nosode made from five homeopathically prepared bacterial sources. Published in January 1986 edition of the *British Homeo-pathic Journal,* this double blind study involved 82 cows, 41 receiving the unmedicated solvent. Results were dramatic—incidence of mastitis in treatment group was 2.5 percent and that of the control group was 25 percent.

The report also includes two other studies. One utilizing *Caulophyllum* 30C tincture added to the drinking water of a herd of Friesian dairy heifers that was experiencing disastrous dystocia problems and the other using the combined mastitis nosode to treat a different herd of Friesian dairy cows with severe somatic cell counts (SCC) of 1,000,000 that had proved unresponsive to all means of control. Both studies produced positive results.

Anecdotal testimony is useful to see how various medicines have been tried and proven in specific cases. For instance, a three-year-old Holstein named Wilma developed an acute, hot and painful right hind mammary gland infection. The secretion was curdy with clots of milk and blood. She resented stripping out and would kick freely. The culture results showed a rapid growth of Pseudomonas, sensitive only to Gentamycin and Carbenicillin. When the owner was advised to sell the cow before she died, the owner replied, "Don't worry about Wilma, Doc, we cured her with *Phytolacca* 30C."

Williamson, Mackie, *et al.,* have been conducting ongoing clinical trials from 1987 through 1990 assessing the use of *Sepia 200C* given postpartum in the prevention of anestrus. Results of several double blind trials indicated that timing may be as important in homeopathic therapy as in conventional prostaglandin treatments. The overall cull rate for infertility problems in this herd was 10 percent, out of which those cows receiving *Sepia* 200C on day 14 had the largest proportion. However, the cows receiving *Sepia* 200C on day 21 postpartum consistently had a higher conception rate with a lower culling rate. The July 1990, July 1991 and January 1995 editions of the *British Homeopathic Journal* contain the complete reports of the studies.

The clinical experience in the United States agrees with these British findings. The point is, there is ample evidence that homeopathy works. This is especially true when it is used as one part of a holistic plan for a healthy farm.

Surgery

Many of the well-known homeopaths of the 19th century were surgeons as well. Experiences in wartime taught them how useful *Calendula* and the other homeopathic medications could be. While I always try to use a homeopathic medicine first, there are times when surgery will be necessary and there are invasive procedures that are routinely performed on farm, for instance, dehorning. Surgery can leave an animal in shock and with considerable distress related to anesthesia and the trauma of surgery. Using homeopathic remedies before and after surgery can improve animal health and reduce healing time.

Homeopathic remedies have value in four areas:
1. Making surgical work unnecessary or limiting its necessity.
2. Preparing surgical cases for operation.
3. Dealing with post-operative care.
4. Symptomatic relief during convalescence.

Medicines used for postsurgical care (with keynote symptoms) are:

Magnesia phosphorica. Patient has crampy pains of abdomen or large muscles, which are better from warmth and application of pressure. Abdominal cramps are often caused by trapped gas in the intestines (*China officinalis* also has a similar post-operative picture.)

Chamomilla. The patient is oversensitive and irritable, post-operative pain is unbearable. Anger and screaming may occur as patient recovers from anesthesia. Animal is worse from both heat and cold, and better from being held or carried while walking.

Hypericum perforatum. Pain along nerves or in fingers, toes (hooves), tails, or ends of teats. Sharp pains after nerve surgery, especially to brain or spinal cord. Patient is worse from jar or touch and better bending backwards.

Hyoscyamus niger. Postoperative sepsis with fever (like Pyrogenium) and excitement. Some post-operative excitement can be avoided by giving pre-operative dose of *Aconitum napellus.* Patient is worse from touch, cold air and at night; he is better from warmth, sitting up and moving around.

Staphysagria. Incision pain and pain in lacerated tissues causes the patient much distress. This is a valuable postoperative medication after hernia surgery and Caesarian section. Patient is worse from loss of fluids and at night; he is better from rest and after a good breakfast.

For non-union of fractures use:

Calcarea phosphorica. Fractures in young subjects. Use after *Arnica* for bone trauma. This medicine is useful for bone pain even before the fracture is stabilized.

Symphytum. From comfrey root; the plant is often referred to as bone-knit. It aids in the repair of non-union fractures that have previously been stabilized. It also helps in healing damaged joints, cartilage and periosteum. Pains are worse stooping, walking and working.

For scar tissue use:

Graphites. Postoperative scarring and cheloids (hard cicatrices); there are non-healing cracks at folds of skin or there is evidence of exuberant granulation (proud flesh) at the incision site. The scars or eruptions often ooze a thick, honey-like fluid. The patient is worse from cold drafts and hot drinks, and is better from wrapping up or bandaging the wound.

Thiosinaminum. Stricture or fibrosis after surgical intervention; fibrosis or scarring of middle ear or structures of the eye or of the intestine, especially the rectum. Deafness due to fibrosis is a common indication for this medication.

"Medical cases may become surgical cases and surgical cases usually become medical cases," stated James W. Ward, M. D. This quote points out how important it is for all medical and veterinary persons to co-operate and work closely for the benefit of the patient. Homeopathy alone is not the solution for each and every patient. Conventional drugs may not be the solution either. There are times when surgery is necessary.

James W. Ward, M. D. (1861-1939) received his medical degree from the New York Homeopathic Medical College in 1883. He taught physiology and gynecology at the Hahnemann Hospital College of San Francisco from 1884 to 1916. Dr. Ward was also Dean of the school from 1899 to 1916. In 1902, he was elected president of the San Francisco Board of Health and became in charge of the relief efforts in the aftermath of the San Francisco earthquake of 1906. He tirelessly worked toward the cause of better health care for all Californians with clinical work, writing and speaking until the day of his death on July 12, 1939. [From *The Faces of Homeopathy* by Julian Winston, pages 126-127.]

Homeopathic medications have been useful at the following times:

1. To help allay fear and anxiety before anesthesia.
2. To medicate surgical sites during a procedure.

3. To reduce hemorrhage and lessen post-operative infection.
4. To smooth recovery from anesthesia.
5. To relieve pain at the incision and speed granulation and union.
6. To relieve general post-operative pain and depression.
7. To reduce scarring and adhesions at the surgical site.

Economic Benefits of Homeopathy

Homeopathic medicines are more economical to use than chemical medicines in both large and small animal practice. In farm animals, this fact is best demonstrated with the symptoms of mastitis. For the purposes of illustration, let us assume that the average cow produces eighty pounds of milk per day and the farmer is being paid $14.00 per one hundred pounds of raw milk. Withdrawal time with antibiotic treatment ranges from three to 15 days after the last dose, depending upon which antibiotic is employed. The following chart compares the economics of three different treatment options.

Treatment Options — Actual Costs

	Milk Withdrawl		
	Antibiotic w/3-Day	Antibiotic w/15-Day	*Bryonia*
Medicine	$15.00	$75.00	$8.00
Vet Call	$35.00	$35.00	$35.00
Milk Loss	$68.00	$168.00	$17.00
Total Cost	$118.00	$278.00	$60.00

In all three treatment options, the medicine is given for four days. In treatment option I and III, the milk is discarded for an additional three days. The difference between those two options is that with the antibiotics, all milk must be discarded whereas with the homeopathic medicine only the abnormal milk from the affected quarter is discarded. The herdsman and veterinarian who uses the well-chosen homeopathic medicine along with probiotics, vitamin therapy, and good management will realize a substantial economic savings. In addition, the patient is healthier after the illness than before.

Generations of Greater Herd Health

Several of our dairy clients have been using a holistic herd health program that includes homeopathy for many years now. Our emphasis has been disease prevention using homeopathic medicines in conjunction with good management and nutrition. Each generation requires less medical intervention than the one before. Calves respond quickly when illness does occur—sometimes only one dose of homeopathic medicine is necessary.

A dairy client related his experience with a two-week-old heifer calf diagnosed with pneumonia. He stated, "The other veterinarian recommended an antibiotic (Naxel), but I wanted to start with a homeopathic medicine first. I gave one dose of *Calc phos* 200C *(Calcarea phosphorica)* because it fit the symptom picture. The next day she was breathing normally, eating well and the fever was gone. I decided I would just watch her closely and not give any drugs. I did not even have to give a second dose of *Calc phos*. She is as healthy as can be."

Those of us who daily care for livestock have a choice to make each time an animal is ill. Will we choose the quick fix or will we work toward long-term health? Our experience is that every time we turn away from the selfishness and expediency of drugs, we are rewarded 100 times over. Practicing homeopathic medicine is work and requires diligence, just as the extra effort made to improve the long-term fertility and health of our soils is work and requires diligence.

If you are farming or practicing veterinary medicine, your long-term goals need to become paramount. One reason for agriculture being in the mess that it is in today is that over the last 50 years the majority have concentrated on short-term goals. As a consequence, the health of the soils, plants, livestock and people has declined.

Keys to a Successful Holistic Operation

1. Take time for observations and good husbandry.
2. Use homeopathic medicines first in every case that applies.
3. Commit to regular herd health examinations with a holistically trained veterinarian. If one is not available in your area, work with your local veterinarian and consult with a holistic vet via the telephone.
4. Keep your medicines clean, properly labeled, in order and preferably in a kit with an instruction booklet.
5. Use vaccines only when necessary.
6. Use conventional drugs and chemicals sparingly or not at all.
7. Have stray voltage checked and eliminated.
8. Have a plan to deal with the unexpected.

There are no cookie cutter approaches on the farms we serve. Although organic farming practices are strongly encouraged, each farm family unit is at a different stage in the journey. Some are moving as fast as possible to become certified while others are on a slower track. A few farmers are taking a cautious wait-and-see attitude, letting the neighbors exercise the pioneering spirit first. Several have worked to establish direct marketing contacts with success. Others struggle with how to set up community sustainable agriculture (CSA) relationships with non-agricultural neighbors.

At this point in time research in veterinary homeopathy is limited. On the other hand, with the increasing number of veterinarians practicing homeopathy and the increasing number of organic farms, there is now a base of individuals and farms from which to collect data that may be developed into appropriate research. And there are educational opportunities in veterinary homeopathy, which have also increased in the last 15 years.

Homeopathy is a discipline that takes time and experience to master, but a farmer can begin here to understand the principles used. One of the wonderful things about homeopathy is that it cannot harm, so even if you make a mistake in remedy choice, your animal will be no worse off than when you began. All it takes is a willingness to try something new and time spent closely observing your herd. Start slow and simple and keep notes on your results. This book will guide you through many of the circumstances faced by dairy farmers and give suggestions for setting up a homeopathic system on your farm. I am convinced that homeopathy is the way to healthier farms, healthier farm families, healthier farm animals and, finally, a healthier planet.

2.

Why Use Homeopathy on the Farm?

Homeopathic medicine allows the farmer to begin with a few basic first aid medicines and then proceed to more and more complex treatments. I recommend that you start with simple first aid cases to build your confidence and knowledge. Many severe health situations can be avoided by following the recommendations in this book. Every well chosen homeopathic remedy will stimulate the immune system of the animal thereby making that animal more resistant to disease and healthier over the long term. A herd that has been treated with conventional drugs will respond to homeopathic medicines if the treatment changes are made slowly over a period of time. It is critical that a period of transition is allowed, but you will see the benefits of homeopathy.

Three points need to be made at the onset. Institute homeo-therapeutics at the first sign of functional disorder. If possible, never give drugs or any other treatments that will suppress a healthy immune response. Always be aware of any side effects of any therapy given. It is always easier and cheaper to prevent disease than it is to treat sick patients. The farmers and veterinarians with whom we have consulted over the last twenty years have become better observers. By the third herd health exam farmers often proclaim, "I am now noticing small changes in attitude just before symptoms appear." This improved ability to observe early changes in attitude (energy or playfulness) leads to a shorter disease course after a homeopathic prescription. In addition, as the veterinarian examines each animal during regular routine herd health exams, the same holds true. Subtle changes like dullness of eye or hair, sluggish activity, irregular estrus cycles, small ovaries, and discolored mucous membranes give the veterinarian reason to institute a homeopathic prescription before clinical disease develops.

> When first using homeopathy it is important to understand that changes take place over time. I know of a case in which an animal was given Naxel, Gentomycin, Albon, Prednisone and Banamine for an illness. As a last resort a homeopathic prescription was given, but the animal did not respond. The strong medicines that had been administered were likely blocking or overwhelming the benefits of homeopathic remedies. This situation frustrated both the owner and the homeopathic veterinarian. In my early years of homeopathic prescribing, I was always amazed when positive results occurred in similar situations. Those few and far between successes kept my interest stimulated to continue trying homeopathic medicines. Now I think homeopathy first and find I generally get good results.

It is always in everyone's best interest to communicate with your local veterinarian regarding your desires to include these medicines in the management of the herd. Changing to homeopathic medicines does not mean you will no longer require the service of a veterinarian. The transition to holistic herd health necessitates a working relationship with an open-minded practitioner. It would be prudent to avoid the use of steroids, anabolic drugs and modified live vaccines with those animals on which you want to use homeopathic medicines. If this advice is not followed, the beneficial effects of the homeopathic medicines will be lessened.

In any emergency situation notify your local veterinarian. Most veterinarians would rather be involved in the case in the beginning than be called after the situation has progressed in intensity.

Predisposition to Chronic Disease

We know that serious illness, allopathic drugs, and vaccination all contribute to chronic disease, but it is also clear that often predisposition to disease is hereditary. Diseases such as heart disease, cancer, diabetes, alcoholism, schizophrenia, and many others have an inherited component. It is also known that DNA, the genetic makeup of an individual, is a factor in predisposition to disease.

In veterinary medicine, it is well known that certain diseases such as Leukosis in certain families of Jerseys, Johne's disease in certain families of Guernseys, heat exhaustion in certain families of Holsteins and others, are clinically observed and documented. Powerful acute diseases such as pasteurellosis, bovine virus diarrhea, strangles in horses and coliform mastitis to name a few can predispose to serious chronic disease if treated with suppressive drugs. If a bovine mother's internal health is compromised, it will be reflected in the health of her calf.

The term *miasma* was first used by Dr. Samuel Hahnemann to described this underlying, hidden disease weakness or the predisposition to disease. Dr. Hahnemann for whom Hahnemann Hospital in Philadelphia, Pennsylvania is named, observed this continuing weakened disease state and only with the use of homeopathic medicines was he able to see an improvement in the health of his patients and thus a lessening of the weakness. These miasmic conditions or chronic disease states exist in livestock and food-producing animals. It takes the form of hidden chronic disease that becomes manifested at critical stressful times of life. The very best farm management, be it organic or intensive rotational grazing or any of the other beneficial holistic farming methods practiced, will see improved health, but chronic disease may still rear its ugly head. It is the wise steward, using the natural/homeopathic approach on a regular basis, who will see less and less chronic disease.

This is an illustration of chronic disease in an organic herd. Good nutrition, particularly that coming in the form of high quality forage, will compensate for some chronic disease in a herd of cattle. It will not eliminate the miasmatic state completely. Any condition that interrupts health and longevity is a *miasmatic* state. In a conventional herd, cows survive about 1.3 lactations. In an organic grazing herd, they last four or more lactations. Those cattle coming into the herd with a heavy weight of chronic disease often begin to show clinical signs during the first or second lactation. One indication is high somatic cell counts (SCC) or frequent flare-ups of mastitis. Another is the failure to thrive syndrome. Certain animals will have thin body condition, rough hair coats or enlarged lymph nodes.

Despite good nutrition, the subclinical chronic disease becomes clinically active disease. These animals need to be treated with the indicated medicine in an aggressive manner. Homeo-pathic medications have the potential to eliminate chronic conditions better than almost any other treatment modality. Proper medication, along with good nutrition, management and handling, will save many of these animals. However, one cannot overlook heredity and if tendencies to take illness is not removed by those actions, the family of cows should be culled.

Support a Healthy Immune System

Getting started with homeopathy involves a change of thinking and a change of action. Conventional approaches to therapy involve killing the bugs with drugs. There are, however, no commonly used drugs that will kill viruses. Homeopathic treatment causes immune stimulation with a simulation. Follow along and I will try to explain simulation, which is more properly referred to as the *simillimum*.

Dr. George Macleod frequently asked this question while teaching, "How does the disease take him?" Regardless of the viral agent present, the symptoms that the patient exhibits continue to be the most important information. Each patient is an individual and responds to stress and environmental changes in a singular manner. Simillimum refers to the total symptom picture of the patient.

Each homeopathic medicine has a symptom picture as well. For example, the keynote symptoms for *Aconitum napellus* are fear, fever and sudden illness with increased thirst. Keynote symptoms for *Gelsemium* are dullness, drowsiness, muscular weakness, slow onset and lack of thirst. When several persons in the same household are suffering with influenza, both symptom pictures may be apparent.

An Ounce of Prevention: Case Studies

Case #1: Farmer A is assembling a herd of Holsteins which he purchased across the state from two different locations. All animals are given *Aconite* before trucking and morning and evening for three days thereafter. No illness occurred and milk production was not depressed.

Farmer B assembled a herd of Holsteins purchased from two different states in the northeast, but was not aware of the uses of homeopathic medicine at the time. He had not yet learned that *Aconite* could prevent illness in animals that are being shipped and stressed. Symptoms of bronchopneumonia and eye infections persisted in herd B for nearly three months; two cows died of the illness, four others were culled, and several lost sight in one eye despite aggressive antibiotic therapy. The expense to Farmer B for lost animals, lost production, medications, autopsy fees, serology and increased labor was nearly $8,000.00. The expense to Farmer A was $40.00 for five vials of *Aconite* plus the time and labor to mix it in the drinking water.

Case #2: Twenty-three dairy heifers aged four to 14 months are brought into the main dairy barn with the lactating herd due to severe winter weather. Fever, with bronchitis and a harsh dry cough developed four days later. Morbidity is near 100 percent. All heifers are treated with *Bryonia alba* in the drinking water and vitamin C in the feed. Two animals with persistent fevers are treated with antibiotics. Five days later 23 heifers are well.

The list of drugs and other substances which have been documented to suppress the immune system is likely to be over a mile long. Any farmer needs to be aware of chemicals, drugs and unhealthful situations that lead to the degradation of the immune system. We need to emphasize that immune suppression does not appear to occur in every patient, but from a holistic perspective these drugs should only be used when there is valid medical indication and generally after trying homeopathic medicines.

Here is a brief list of some immune suppressive drugs and conditions:

- Cortisone and all steroids
- Preservatives in feed
- Live and modified live vaccines
- Antibiotics
- Antihistamines
- Drugs used to prevent and control parasites
- Elevations in ammonia in barns and stables

Animal caretakers must weigh each therapeutic decision carefully. When in doubt, always err on the side of stimulation rather than suppression. Homeopathy, used in conjunction with good herd nutrition and care, can boost the immune system. Following are a few ideas to support the immune systems of your animals:

- Energy medicines like homeopathy and acupuncture
- Vitamins and Minerals
- Probiotics
- Exercise, Massage, Hydrotherapy
- Raw foods and whole grains, organic if possible
- Daily exercise year round
- Feed good quality forages—both fresh and baled

Know the Possible Results
of any Therapy: A Case Study

The Clark Veterinary Clinic was called to visit a calf rearing facility where the mortality rate on the entry group exceeded 50 percent. Calves were being purchased at three to seven days of age from surrounding dairy farms and housed in large box stalls or calf hutches. Illness began at one to two weeks of age. Death was preceded by gastrointestinal or respiratory symptoms lasting one to three days. Previous veterinary advice consisted primarily of vaccinating with a modified live combination injectable vaccine (MLV) as early in life as possible. Prior to instituting the vaccine, mortality was 20 percent; after the MLV vaccine, mortality exceeded 50 percent.

The owner/manager needed to become knowledgeable about the differences between the various therapies available. It was obvious that the calves were not being helped by the MLV vaccine. If only a few animals in the herd were being saved, should the vaccine be continued? Is there an-other way to stimulate (or boost) immunity? Will antibiotics help? Several different antibiotics had been used over the preceding months with no favorable results. In fact, the one positive change that had been made was to stop using medicated milk replacer.

The question which we asked was this: "Would you be willing to stop all conventional drugs (vaccines and antibiotics) in favor of natural therapies that will stimulate a healthy immune response? Bear in mind that vitamin C, probiotics and homeopathic medicine will not directly kill viruses or bacteria."

Fortunately, once all the facts were presented, the decision was made to proceed with therapies that will stimulate rather than suppress. Homeopathic *Aconite, Pulsatilla* and *Phosphorus* were prescribed over the following weeks with good results. In addition, an oral nosode was prepared from the most ill patients. This was used to protect each new arrival in the future. (A nosode preparation is made from a small amount of the offending bacteria or virus.) Mortality dropped to less than two percent. As natural therapies were continued, a healthier herd of calves emerged and economical benefits were realized.

Making the Nosode

While on the farm I examined 40 or 50 calves. Ten of the most seriously ill calves were set aside for sampling. Generally, just before an animal succumbs to an infectious disease it is shedding viruses or bacteria from many body openings. We collected mucous from the nose, throat and colon using sterile swabs. The mucous from the swabs was used in the preparation of a new homeopathic medicine. I asked the farmer's wife if she had any whiskey, gin or brandy. She brought an old bottle of brandy up from the cellar.

The dust on that blackberry brandy was thick. We wiped the bottle as clean as possible and used the brandy as a source of ethanol for each stage of the manufacturing process. An old oak table became our pharmacy table, and a thick telephone book served for succussion. An hour later the homeopathic nosode had been created in 12C potency. I remember that the entire house smelled like blackberries.

All of the most critically ill calves exhibited green, creamy discharges. They were dosed with *Pulsatilla* 30C and the new medicine (12C) three times daily for four days. Four days later, all calves on homeopathic medication were recovering. Not one was lost.

Bovine Calf nosode 12C may be dosed along with the most indicated homeopathic medication in the treatment of various respiratory diseases. To aid in prevention of these conditions, dose new arrivals with *Aconite* and then with the nosode a.m. and p.m. for three days. Follow-up dosing with the nosode once weekly for five weeks may be advised for certain farms.

What Do I Need to Begin to Use Homeopathy?

If you are reading this book, then you have already started to learn about homeopathy. There are many books available on the subject, *The Treatment of Cattle by Homoeopathy* by Dr. George Macleod is one such text. This book sets the modern standard by which all other works may be compared. I have included a brief list of these books in the References section. I recommend reading and studying homeopathy as you have time, you will be amazed at how quickly you can pick up the skills necessary to successfully prescribe some homeopathic medications.

A *Materia Medica* is a comprehensive list of homeopathic medicines and nosodes. There is one in this book that is particularly focused on the bovine species. Another one I often use is *Materia Medica with Repertory,* written in 1927 by William Boericke, it is a classic that has been used for generations as a reference. Over 1,000 pages in length, the book lists many medicines for man or animals in both common and Latin names. The most important thing is to find the reference that is easy for you to use and then use it.

A homeopathic kit is one that everyone should have in their home or barn. There are many good home kits on the market. Perhaps your local health food store or pharmacy has one that they recommend. All of our clients are encouraged to purchase a kit which should come with a large number of basic medicines and a guide to their usage. Asterisks in the chart below denote medicines that I believe should be in any basic kit (the uses of and indications for these medicines can be found in the Keynotes for Homeopathic Medicines chapter):

Basic Homeopathic Remedies

	Pellets	Dilution	Ointment	Gel	Lotion	Spray
Aconitum napellus	*	*	*			
Antimonium tartaricum	*	*				
Apis mellifica	*	*	*			
Arnica montana	*	*	*	*	*	
Arsenicum album	*	*				
Belladonna	*	*				
Bryonia alba	*	*	*			
Calcarea carbonica	*	*				
Calendula officinalis	*	*	*	*	*	*
Carbo vegetabilis	*	*				
Caulophyllum	*	*				
China officinalis	*	*				
Echinacea/Rudbeckia	*	*	*			
Ferrum phosphorica	*	*				
Gelsemium sempervirens	*	*				
Graphites	*	*	*			
Hepar sulphuris calcareum	*	*				
	Pellets	Dilution	Ointment	Gel	Lotion	Spray
Hydrastis canadensis	*	*				

Hypericum perforatum	*	*	*	*	*
Iodium	*	*			
Ipecacuanha	*	*			
Lachesis	*	*			
Ledum palustre	*	*	*	*	
Lycopodium clavatum	*	*			
Magnesia phosphorica	*	*			
Mercuris corrosivus	*	*			
Nux vomica	*	*			
Phosphorus	*	*			
Phytolacca decandra	*	*	*		
Pulsatilla nigricans	*	*			
Pyrogenium	*	*			
Rhus toxicodendron	*	*	*	*	
Ruta graveolens	*	*	*	*	
Sepia	*	*			
Silicea	*	*	*		
Sulphur	*	*	*		
Symphytum	*	*	*		
Thuja occidentalis	*	*	*		
Urtica urens	*	*	*	*	

In addition to reading, study and using your homeopathic kit, there are conferences, courses and seminars that you can attend. Each month *Acres U.S.A.* lists upcoming events in a state by state directory and the American Holistic Veterinary Medical Association annual meeting is held in the month of September at various locations throughout the United States. A little time spent on the Internet will turn up more information than you will be able to use.

First Aid

First aid cases are always excellent opportunities to incorporate homeopathy into the herd and learn their effects. For instance, *Arnica* is the medicine of choice in trauma in all its varieties—mental or physical—and their effects recent or remote," states Dr. Walter Schollaert of Belgium. According to Dr. Schollaert's article in the *IAVH Newsletter* of June 1997, "the anti-inflammatory and analgetic properties of *Arnica* can be explained by inhibition of the enzymes involved in inflammations. The active components of the *Arnica* plant contain sesquiterpeenlactones which play a role in the inhibition process."

Slipping on wet concrete or giving birth to a large calf can produce a nasty injury in a cow. Damage to the obturator nerves as well as tearing of muscles and ligaments can lead to paralysis. The first medicine of choice should be *Arnica,* dosing three or four times daily for three days. As function returns, *Hypericum* may be given three times daily to speed neuromuscular healing. If any abnormal gait persists the following medications often complete the cure: *Gelsemium* when groups of muscles are weak; *Ruta graveolens* when joints are still inflamed, and *Magnesia phos* when cramps in muscles are the principle disabling factor.

On a farm, it is good to develop a game plan for prevention and treatment. One part of that plan is learning and using homeopathic medications. Homeopathic medicines have been safely prescribed in animals for over 200 years. In Europe and in North America, customer demand for organic foods continues to outstrip supply. Homeopathy is a system of medicine that can easily be learned by organic farmers and veterinarians alike. It is never too early to begin the educational process.

The best way to learn, as with any other discipline, is by personal experience. Using the medicines described in this book on the farm will forever set it in your heart and mind that homeopathic medicines are valuable tools. One way to begin is to learn homeopathic medicine by studying one medicine at a time. Choose a medicine to begin with and then read about it in the texts you have on the farm. Once you are familiar with a medicine, begin to use it.

Arnica montana is an excellent medicine with which to begin because it is used in so many different situations. *Arnica* is available in oral tablets and pellets, topical ointments, lotions and sprays. As shown above, it can be used for any number of ailments, here are a few more:

1. Tramped udder and teats. If the area is lacerated, use *Calendula* ointment or gel topically. If only bruised, *Arnica* ointment, cream or lotion may be utilized.

2. Use following any surgical procedure such as castration, dehorning, neutering or spaying, debeaking or emergency surgery.

3. *Arnica* is useful for difficult birthing in all species.

4. Bruising-haematoma. Oral and topical application may be appropriate after trimming and shoeing or any situation in which bruising occurs.

In all the above situations, *Arnica* 12X, 30X or 30C could be administered three times daily for three days. Decrease the frequency as the animal improves. The number of tablets or pellets to give depends upon the size of the mouth of the animal in question. A Yorkshire terrier would receive two to three pellets or tablets as compared to a Percheron gelding who should be given eight to ten pellets or tablets. When medicating small animals such as pet birds or rabbits, one pellet dissolved in the daily drinking water is very effective. When animals are severely ill, spraying the medicated water on the face or in the mouth works well. It is always best to give homeopathic medicine at a separate time from food or other medicines.

Once you have a working knowledge of one homeopathic medicine, move on to another. In no time you will have a large repertoire of knowledge to draw from whenever a potential health problem arises.

3.

A Holistic Operation

Each holistic practice is unique, offering the consumer a variety of healthcare options depending upon the interests and expertise of the practitioner. A common thread running through all of them is the emphasis on the "whole" patient. No longer is it just a blocked urethra or a fracture case or a case of chronic arthritis. Instead, each animal and each case has its own particular holistic requirement. It will be the choice of the practitioner and client to decide the treatment regimen.

In bovine veterinary homeopathy, the cow is a member of a herd and the herd lives on a farm which in itself is a dynamic living entity. The herd affects the farm and soil, the soil affects the herd and farm and the farm affects the herd and soil. A holistic bovine veterinarian looks at the whole and the parts of the whole knowing that each is playing a role in the total health picture. Each influences the other and in so doing produces a picture of health and/or disease. A practitioner may either choose to prescribe for the individual cow or the herd or to make recommendations for the farm. Whichever holistic approach is chosen, the results will likely be improved herd health.

Homeopathic medicine:
- is safe for both the environment and the farm
- produces no adverse side effects
- causes no immune suppression
- increases the vitality of the succeeding generations
- provides an option for the treatment of chronic diseases
- is cost effective
- produces no drug residues

A farmer can sleep peacefully knowing that the milk in his/her bulk tank contains no forbidden drugs. Once the practice of first using natural medicines is adopted, there is rarely a need to administer drugs with withdrawal times longer than zero days. A farmer can relax knowing that he has a contingency plan at any time a crisis could occur. A farmer can know that his children are safe from the accidental ingestion of chemical drugs and that his wife and daughters are protected from the possible side effects of hormones like BGH (bovine growth hormone). A farmer can rest easy knowing that, by eliminating chemicals on his farm, he will not be contributing to the possible infertility of his sons and daughters. A farmer can save money knowing that his veterinary drug bills will not be a cumbersome burden.

When examining a herd, my first question is "Are these contented cows?" Some years ago an advertisement encouraged consumers to "drink milk from contented cows." A contented, peaceful cow will, in all probability, be producing a quality product. An ill or diseased animal will show signs of stress along with the illness and will be producing an inferior-quality product, low in nutrients such as CLA. (CLA is an abbreviation for Conjugated Linoleic Acid, a nutrient discovered in the 1980s which has many health benefits. Traditional diets that include raw milk, raw milk butter, red meat and organic lard are high in this wonderful nutrient. Conjugated Linoleic Acid is highest in foods from livestock grazing outdoors on fast-growing grasses and other plants.) Because of what they eat, the exercise, fresh air, health care and sunshine that are such an integral part of the daily lifestyle of holistic operations, milk and meat from these 'contented cows' are more nutritious.

Homeopathic medicine may be used in the following three main situations: to treat ill animals, to prevent illness and to maintain health. In all three areas drug residues are eliminated. With the animal's immune system in balance, invading bacteria or viruses appear to have little effect. And if the animal does succumb to the invading organism, they appear clinically to show a strong and rapid response with a return to homeostasis or stable health status within a short period of time.

Quick Return to Health: A Case Study

A calf developed symptoms of pneumonia. The client had the choice of antibiotics or *Calcarea phosphorica*. He chose *Calcarea phosphorica (Calc phos)* because in previous years his calves on that same farm responded well for symptoms that included weak and crooked legs, painful dentition, diarrhea, and respiratory affections. The calf showed marked improvement within 12 hours. This particular client has chosen to use natural medicines and has been doing so for eight years. His cattle are not suppressed by overuse of drugs and vaccines, so they generally respond quickly to energy medicines.

Vaccinations

To vaccinate or not to vaccinate is always the farmer's dilemma. With the natural approach, you have additional options. Determine the underlying disease problem in the herd by using serologic testing for BVD, IBR, Lepto, Haemophilus, etc. If any of these are present in the herd, you may choose to use the killed injectable vaccine for those specific disease agents followed by the corresponding homeopathic nosode. Or you may choose to just use the homeopathic nosode for those specific disease agents. As prescription-only medicines, your attending veterinarian would be able to supply those nosodes to you and in turn monitor the herd following the administration of them. Clinically, it appears that a herd treated in such a way maintains and improves health. Individual animals have longer, more productive lives when they are not over-vaccinated.

Homeopathic scientists have long believed that placing undo emphasis on vaccinations to prevent and control disease is flawed. During the great epidemics of yellow fever, cholera, typhoid, and influenza, they employed hygiene, quarantine and nutritional measures along with the potentized medicines to achieve success. They learned from experience that attempts to immunize with vaccines frequently were ineffective and also that patients that had been subjected to faulty vaccination programs suffered from new forms of chronic disease.

Facts About Vaccines

Victor S. Cortese, D.V.M., the managing veterinarian of technical services (cattle) for Pfizer Animal Health, states, "Stress and nutrition are very important for a healthy immune response. No vaccines should be given during the late stages of pregnancy and for seven days after labor and delivery. Animals should not be vaccinated if the environmental temperature is above 90 degrees Fahrenheit. The micro-minerals copper, iron, zinc and selenium, and vitamins A and E must be present (and in the correct proportions). Vitamin E is important in all components of the immune system and in protecting animals from oxidative damage. Leukocytes contain 10 to 35 times more alpha-tocopherol than do red blood cells or platelets. The immune system is a complex and interrelated system. Vaccination does nothing more than try to trick the immune system into responding as if it had been attacked by that disease. The closer a vaccine is to the natural wild virus, the stronger the immune response will be, but greater also are the chances of causing disease. Vaccines that completely stimulate all aspects of the immune system should be selected."

Neonatal Immunology

According to Dr. Cortese, "The field of neonatal immunology is going through a revolution both in human and veterinary medicine. New and advanced methods of assessing immune status and function are indicating that an important component of the newborn food-animal's defense mechanism is colostrum. Constituents of colostrum include: concentrated levels of antibodies, many of the immune cells (B cells, CD cells, macrophages and neutrophils) which contain interferon, an essential immune system chemical, and many nutrients in concentrated forms. Since digestive tract absorption of the colostrum nutrients decreases within hours of birth, the quality, quantity and timing of colostrum administration is important."

"Stress impacts the neonate's immune system as it does older animals. Factors that can affect the immune system are the birthing process and the presence of increased amounts of suppressor T cells during the first week of life. Systemic vaccination during this time should be avoided due to these decreased responses and may even have undesired effects. Other stresses such as castration, dehorning, weaning and movement need to be avoided in the immunologically frail newborn."

Adverse Reactions

In veterinary practice, adverse reactions, severe side effects and immune suppression from nearly all the vaccines are observed. Jean Dodds, D.V.M., noted authority and researcher in immunology and thyroid disease, states, "Vaccinating animals at the beginning, during or immediately after an estrus cycle is unwise as is the relatively common practice of vaccinating animals during pregnancy or lactation. . . .Recent studies with MLV herpes vaccines in cattle have shown them to induce necrotic changes in the ovaries of heifers that were vaccinated during estrus. The vaccine strain of this virus was also isolated from control heifers that apparently became infected by sharing the same pasture with the vaccinates." Dodds also reports immune reactions following MLV canine parvo virus vaccinations. Signs and symptoms observed have included fever, stiffness, sore joints, abdominal tenderness, collapse with auto-agglutinated red blood cells and icterus, generalized petechial and ecchymotic hemorrhages, liver and kidney failure, bone marrow suppression, transient seizures, bloat and stained teeth.

New Recommendations Regarding Vaccines

Larry Glickman, V.M.D., professor of epidemiology and environmental medicine at Purdue University School of Veterinary Medicine, recommends, "Instead of using combination vaccines, veterinarians could one year give the necessary distemper vaccine and give the parvo vaccine the next year. Alternating vaccines so they use only one vaccine a year is a reasonable solution." In *Veterinary Product News*, it was reported that a foundation that provides funds for research in childhood genetic diseases approached Glickman with the proposal to use dogs as a model to study vaccines and disease. The two-year funding has led him to conduct experiments that mimic what would be done in practice. He is looking at the effect of a single course of immunity repeated over time. His team is using multivalent vaccine for canine distemper, parvo and rabies. "We think the reactions we're seeing are due to the contaminants in the vaccines. The viral and bacterial components grown in culture systems for vaccines are supplemented with bovine serum to encourage the growth of organisms. But when the viruses are grown and put into the vaccine, those cow proteins are not removed. Not only are we immunizing dogs and cats and humans against viruses, we're immunizing them against the cow proteins and there are hundreds of them. That is not what we want," said Glickman. His proposed solution: use serum from the species being vaccinated.

Glickman's recommendations are supported by Dr. Ronald D. Shultz, Ph.D., chairman of the Department of Pathobiological Sciences at the University of Wisconsin, Madison, School of Veterinary Medicine and Dr. Dennis W. Macy of Colorado State University. Giving one vaccine at a time allows the immune system an opportunity to respond to one antigen (or one protein) at a time. The multivalent vaccines have too many components, giving rise to both acute and chronic adverse reactions. According Dr. Shultz, "No one ever expected to take a normal healthy cat in for vaccination and end up with terminal cancer. Even worse than that—prior to knowing about fibrosarcomas, you could take a normal healthy dog in for its shots and end up with a dead dog (because of adverse reactions). "Both the University of Minnesota and Colorado State University are now recommending that dogs receive only one vaccine a year. "What we're concerned about, of course, is that we don't want to do any harm," says Dr. Macy. "We have had increasing evidence that the duration of immunity in many of the products is much longer that once was perceived."

Immunization Protocol for Organic Dairies

In herds of animals, especially cattle, multivalent modified live vaccines have been routinely administered when a contagious disease has been suspected of causing illness. Until recently the recommendations have been to vaccinate with an eight-way, nine-way or even a ten-way vaccine at the same time. Most organic farmers believe that multivalent injections cause undue stress to the individual and to the entire herd. Homeopathic veterinarians dose with one medicine at a time and tend to use vaccines in the same manner. Thankfully, today's research has validated the cautious approach of organic farmers and homeopathic scientists.

One practice that appears to be beneficial is to follow the vaccine with the corresponding homeopathic nosode. When herds are vaccinated, the nosode can be administered in the drinking water. In herds on a holistic or homeopathic health program, nosodes alone may be used in lieu of vaccines.

Management, cleanliness and hygiene should not be taken for granted. Stalls, pens, trucks and trailers should be thoroughly cleansed of bedding, manure, urine and all other body discharges. In most cases, steam cleaning or high pressure hot water works well. Thereafter, stalls should be disinfected and allowed to air dry for at least seven days. New arrivals should be quarantined (isolated) for a minimum of 30 days.

No one should ever depend on any vaccine alone to prevent disease outbreaks. Animals given live or modified-live vaccines always possess the ability to shed virus if they are crowded, stressed or treated with corticosteroids. Vaccinates housed with non-vaccinates have potential to spread contagion. Animals given killed vaccine may become ill from the chemicals (in the bottle) used to inactivate the organisms. Immediate administration of the corresponding nosode for one to three days after the vaccine may lessen side effects.

Dr. Don Sockett, veterinarian and virologist with the Wisconsin Department of Agriculture, Trade and Consumer Pro-tection, recommends focusing on four general areas: isolation, testing/

monitoring (to find and follow the disease active in the herd), vaccination, and sanitation. In a recent issue of *Dairy Today*, he states, "A complete vaccination program for both your existing herd and new animals is probably the simplest, most cost-effective step you can take to head off a disease outbreak. But you also have to remember that just because you vaccinate, it doesn't mean animals are protected."

If animals are to be vaccinated, we should attempt to correct the following factors first:

Nutrition: is it balanced in mega nutrients? Is the feed adequate in micro nutrients? Leukocytes need tocopherols (vitamin E), trace minerals and natural antioxidants.

Stress: are housing and handling contributing factors? Are animals being shipped? Do not vaccinate twelve days before shipping or for seven days afterward. Avoid vaccines during early pregnancy, late pregnancy and estrus, and for seven days after giving birth.

Maternal antibodies: in most species these are present until five months of age, and longer if the neonate continues to nurse. Therefore, live vaccines given before five months of age may be neutralized.

Homeopathic medicines and nosodes: are you treating according to the simillimum? Are you using nosodes from your herd, or for the disease that has been diagnosed in your herd? If we must administer a vaccine, remember to inoculate for only one disease at a time whenever possible.

All animals regardless of species are truly like the canaries in the mine shaft. What is true for them is also true for mankind. Quality health is comprised of many things including wholesome organic foods, adequate housing and humane husbandry, a low-stress lifestyle, and cautious use of available vaccines.

The Monthly Herd Check

It is always easier and cheaper to prevent disease than to treat sick patients. For those farmers who are contemplating becoming certified organic, start to change the health protocols for your livestock as you make those necessary adjustments in the health of your soil. Making simple changes can easily be accomplished with the wide variety of homeopathic medications and products that are available. The initial cost will be minimal as only a few products need to be purchased and utilized into the operation. Farmers and veterinarians alike will improve their powers of observation as the medicines produce results, and confidence levels rise. If homeopathic medicines are used following the regular herd health examinations, the farmer will begin noticing small changes in attitude (energy and playfulness) just before abnormal physical symptoms appear. This improved ability to observe early changes in attitude leads to a shorter disease course after homeopathic prescriptions.

The homeopathic/holistic approach to a monthly herd evaluation or herd check requires the examining veterinarian to use all of his or her senses not just the brain or the arm. From the moment the veterinarian steps out of the vehicle and onto the farm, he or she is collecting data. A homeopathic farm practitioner learns to practice by observing the whole operation rather than just the individual cow. Observing the whole operation is a developed skill. The homeopathic veterinarian or practitioner is listening and observing and thinking all at the same time. Observe the condition of the buildings, fencing, milking facility and equipment, lighting ventilation and the pastures. All of this will influence the status of the herd.

Listen to all the facts. The farmer may assume that his problem is just the mastitis, but many factors lead to the manifestation of that problem. A holistic or homeopathic approach precludes any assumptions before all the data is collected. Each operation is different.

Look at the whole cow—nose to nates or tip of muzzle to tip of tail. The examination of the herd is more than "doing a rectal." A healthy cow should be contented, with a bright, clear eye and smooth hair coat. A nervous, agitated demeanor, or inflamed eyes, or a runny nose, with tenderness over the liver points signals that there is some sort of "dis-ease" present. Prodromal signs of illness indicates subclinical disease, which will often become clinically manifested. The physical exam is of most importance, next is the history, followed by nutrition and environmental influences. Important aspects of the internal and external exam should be enumerated and the dairyman or assistant should record all the findings.

When examining the reproductive tract, let it tell you what the cow needs. The following describes four typical states of a dairy cow. Treatment consists of dosing with a 30X or 30C once or twice daily for one week. Timing of dosing is very important, and the presence of a CL (see box below) helps to determine the best time to begin the medication.

A "CL" or corpus luteum or yellow body is a progesterone secreting yellow glandular mass in the ovary that is formed from the follicle wall after ovulation. In pregnant cattle, the corpus luteum helps to maintain and nourish the early developing fetus. In non-pregnant animals, a persistent CL will delay the next estrus. A cystic CL will show a similar clinical picture. A veterinarian can see these structures on ultrasonic examination. In the past, veterinarians were able to identify these ovarian structures via rectal palpation. Abbreviations: CLRO = corpus luteum right ovary, YBLO = yellow body left ovary.

The *Sepia* Cow: cervix feels hard, long, broad or enlarged. The anterior part will be hanging over the pelvic brim. Overall the cow looks like she has had one too many calves. It is best not to breed for one or more estrus cycles while healing is taking place. Animals that have given birth annually for four or more years often need a time of rest and recuperation between pregnancies. *Sepia* given after day 21 postpartum results in improved conception rates and shorter calving intervals.

The *Pulsatilla* Cow: in the early part of proestrus the follicle is developing. Overall the cow looks in good flesh, but may have a creamy discharge as she enters estrus. Any concurrent symptoms will be worse from a stuffy barn, and better outside in the open air. Her personality is of a friendly timid cow that is most comfortable with the herd rather than being alone. Dose a.m. and p.m. for three days.

The *Silicea* Cow: the cervix is scarred, one feels a knot and the scaring can sometimes be into the vagina. Other concomitant symptoms might include inability to pass the insemination rod, or frequent services with no conception. The body type tends to be thin and weak in fiber. Dose once daily for seven days, then twice weekly for one month.

The *Apis* Cow: her ovaries are cystic and she has frequent estrum or is very active. There may be air in the vagina and the cervix is inflamed. Usually there is a cyst on one or both of the ovaries. Dose with *Apis mellifica* 30C twice daily for three days, followed by *Natrum muriatricum* 200C once daily for seven days.

For non-lactating cows with no mastitis problems, *Calcarea carbonica* and *Silicea* mixed together in the drinking water every 14 days during the dry period has been helpful. The cows respond with fewer cases of fresh cow mastitis than in previous years. In addition, after a year on the program, the herd somatic cell counts (SCC) will tend to stay at lower levels.

With the potentiality that BVD, Leukosis and Johne's disease can be transmitted using a common rectal sleeve, it is prudent to change the sleeve with each patient. Research performed in Scandinavia and at Washington State Veterinary College have confirmed that disease transmission by common sleeve is a definite possibility. The same precautions should be observed when using needles, syringes and tattooing equipment.

Common Dosing Abbreviations
SID = once daily
BID = twice daily
TID = three times daily
QID = four times daily
EOD = every other days

Walk your pastures and check your barns and stalls regularly. Obstacles to cure prevent the immune response even after the correct prescription has been made by the veterinarian or practitioner. Homeopathic medicines work best in an organic program where fewer obstacles exist. Some of these obstacles are not always apparent such as:

1. Poor ventilation in barns and stables;
2. Lack of adequate bedding;
3. Noxious ammonia fumes from gutters and manure pits;
4. Low carbon and organic matter in fields;
5. Feed containing blood meal, animal parts, and urea;
6. Overuse of unnecessary multiple vaccines;
7. Indiscriminate use of steroids, antibiotics, and hormones; and
8. Uncontrolled current or stray voltage in the milk house, milking parlor, barn or pastures.

Naturally Reared Calves

Calves need a diet of milk and whole grains (oats, barley, spelts) or grass. If a milk replacer is used it will need to be free from antibiotics, animal fat, animal protein factor, artificial flavors and colors. They may have milk protein and plant and grain protein, but the best milk replacers are those made entirely of dried milk and milk protein. It is always preferable to leave the cow and calf together as long as possible.

There is some anecdotal evidence that when young animals are fed the animal protein by-products, they will be unable to digest complex protein later on in life. A veal producer or farmer raising calves for replacement stock in his herd may want to spend an increased amount for a calf milk replacer product without artificial flavors, colors and preservatives and animal by-products. Or he may choose to feed them whole milk (raw) or even whole milk mixed with on-farm produced colostrum. You only get what you pay for. A cheap price may mean inferior quality.

If you are raising calves for veal, grazing is best. A high quality veal is produced when calves are grazed on young succulent forage. In most years veal calves will thrive on milk, a little hay or grain, and lots of grass from April through October. In winter or during a drought, good quality mixed hay is necessary to insure development of a healthy gastrointestinal tract.

Raising calves without antibiotics using natural methods is possible and will produce the best animals over time. Having the homeopathic medicines readily available is a prerequisite to success.

Effects of Poor Quality Milk Replacer: A Case Study

A farmer had eight Holstein calves, robust and healthy at birth, but which shortly thereafter developed low-grade pneumonia with loss of weight. Even though he treated them with *Phosphorus* and *Bryonia,* he suffered a mortality of 25 percent. The remaining six calves continued to eat large amounts of milk replacer and some grain and hay, but did not improve in health. Upon physical examination, the hair coat was fine and thin, and the bodies of the six survivors appeared devoid of external fat. The gums did not appear healthy and they continued to lose weight, looking cachectic.

The package label for the milk replacer indicated the presence of 20 percent animal fat (tallow, lard, etc.). A newborn calf's gastrointestinal system cannot digest or absorb animal fat or animal protein. Since the farmer had just bought a supply of the product a compromise was reached. He would feed the calves one-half milk replacer mixed with half whole raw cow's milk and administer *Calcarea phosphorica* 30C once a week until symptoms began to improve.

Additives in Feed

Nutrition is one of the pillars of a solid farming foundation. Now with the publicity of animal by-products in the feed, it is even more important for the producer to communicate with his local feed company to be certain that his ration is free from those animal by-products known in the industry as bypass protein of animal origin, tallow and animal fat. Sometimes these ingredients are indicated on the label as animal protein factor or protein hydrolysate.

One of my clients with a persistent SCC problem was encouraged to eliminate the animal protein from his feed. Almost immediately his production per cow increased and with homeopathic treatment his SCC problem has also improved. However, even in this case, until the pastures and tillable land have been replenished with nutrients, the SCC problem will not be totally under control. A holistic approach to animal health must include healthy soil and crops.

Odors on your Farm

According to Joel Salatin, "If you smell manure, you are smelling mismanagement." On commercial farms it is not difficult to smell manure. Odors can be controlled by increasing the carbon in the bedding, by using soft rock phosphate in the gutters, and by applying a homeopathic spray. A homeopathic spray developed on the east coast has been useful in box stalls, gutters, holding tanks and run-in sheds. In the end, there is no odor treatment that works as well as keeping barns and farmland clean and free of debris.

Economics

Economically, over a twelve-month period, a producer should see a decline in medical costs for the herd when the switch is made to a holistic operation using homeopathic medicines. As the producer becomes more knowledgeable regarding the natural medicines and how to use them, the need for more expensive chemical drugs should decrease. I remember Roman Stoltzfoos' presentation at the 1995 Acres Conference, "I've had one emergency vet call this year." With his use of natural medicines on a routine basis, his cattle are healthier and therefore less likely to develop those emergency conditions that would require a veterinarian. For the veterinarian, a fire engine practice is gradually replaced by one of preventive care. As a regular client told me just the other day, "Since we are rotating our pastures and using homeopathic medicines in our program, none of the diseases that troubled my herd three years ago seem to be around anymore."

Skeptics Become Believers

Of all alternative and complementary medical therapies, homeopathy probably engenders the most emotional debate. Understandably, conventionally trained clinicians balk at the central premise of homeopathic medicine: a method of treatment involving therapeutic use of solutions containing substances diluted to the point at which they contain virtually none of the original solute. However, analysis of study results indicates that homeopathic products may exert a physiologic effect. Homeopathy is defined as the treatment of animals that have signs of disease with a specially prepared substance that, in sufficient quantity in a healthy animal, would cause clinical signs. Susan Wynn, D.V.M., "Studies On Use of Homeopathy in Animals," in *Journal of The American Veterinary Medical Association,* March 1998.

The exact mode of action of a homeopathic medicine is unclear. It is this fact and this fact alone that will probably remain the stumbling block to the acceptance of the medical discipline of homeopathy. But if something works over and over again, then it seems to me that experience is saying it does work. The experience by farmers and homeopathic veterinarians in practice confirms this idea. For both the short term and long term, the results are de-monstrable, verifiable and repeatable. A specific symptom picture in matching patients will respond to the simillimum again and again. Each time the correct prescription is given, similar results are observed.

For example, *Sepia officinalis* is a frequently used homeopathic medicine in dairy herds. The keynotes of *Sepia* may include decreased tone of uterus, large and saggy udders, loose ligaments, tendency to prolapse uterus, hind leg weakness, infertility, discharge after calving, poor digestion, reddish urine, yellow color of muzzle, lack of affection for the calf and, in general, an animal that's worn out and tired. They tend to be better from warmth and exercise; worse from cold, before a thunderstorm and in the evening. When more than three or four of the above symptoms are observable in a dairy animal, then *Sepia* may be prescribed with confidence. Repeatable results will occur. Again and again, *Sepia* has proven to be a reliable choice.

Monitoring progress of homeopathy in a herd of cattle can be accomplished as a whole group or individually with the use of Dairy Herd Improvement Records (DHIA), somatic cell counts, rolling yearly herd average, etc. Additionally, maintaining a database for each individual cow on each individual farm can, over time, provide usable information regarding whether or not the homeopathic principles are demonstrable, verifiable and repeatable. Facts are facts. Cows do not lie. They either produce milk and calves or are culled.

A Tale of Two Herds

Individual cows in the following two herds were chosen at random to illustrate diversity of symptoms and diversity of remedy choice. These are neither the best cows in the herd nor the worst. This is merely a progress report of these individual cows within these two different herds over the past two years.

Herd "A"

Herdsman "A" began farming three years ago using only homeopathic medicines from the first day of operation. He utilized general grazing the first year and intensive rotational grazing the second and third years. Presently, he is transitioning the herd toward seasonal calving. Cows that calved in the summer and fall of 1997 were not bred again until May of 1998. Those that calved in the spring of 1998 were bred back by June of the same year. Breeds include Holsteins, Jerseys, Milking Shorthorns, and the ever popular Jersteins.

In August of 1996, Dorothy, the Holstein cow, was treated with *Causticum* 30X for ten days for retention of urine with involuntary passage of urine when walking, and decreased pelvic and bladder tone. Would the general pelvic weakness lead to a miscarriage? No other medicines were given and no surgery was performed.

> **Dorothy**
> bred 6/15/96 (age 4)
> calved 3/30/97
> checked 5/16/97—CL (yellow body), good tone.
> Rx: *Calcarea carbonica* (a.m.) and *Pulsatilla* (p.m.) for
>   seven days.
> bred 6/19/97
> checked 9/16/97—pregnant
> calved 4/4/98

Alice

calved 7/17/97 (age 3)—dairy type with good
   production.
checked 7/22/97
Rx: *Calcarea phosphorica* daily for ten days.
bred 9/20/97
checked 11/25/97—pregnant
calved—6/98

Nite

bred 10/8/96 (age 8)
checked 11/26/96—pregnant; painful pelvic organs;
   history of miscarriage
Rx: *Cimicfuga racemosa* once daily for ten days.
calved 7/17/97
checked 7/22/97— very active and excitable.
Rx: *Phosphorus* twice daily for three days, then *Sepia*
   twice daily for three days. Plan to breed in May of
   1998.
checked 4/28/98—sagging uterus with urine pooling.
Rx: *Causticum* once daily for ten days.
checked 5/29/98—better tone with no urine found in
   reproductive tract; follicle left ovary.
Rx: *Ovarian* twice daily until estrus.

Holly

bred 9/30/96 (age 6)
checked 11/26/96—pregnant
calved 7/1/97
checked 7/22/97—scarring of cervix.
Rx: *Sepia* once daily for a week, followed by *Silicea*
   once daily for one week.
checked 3/5/98—open.
Rx: *Sepia* twice daily for three days followed by
   Pulsatilla twice daily for three days. Will breed in
   May or June to calve in February or March.

Herd "B"

Having farmed for nine years, herdsman "B" began using homeopathic medicine nine years ago. This operation is exceptionally well managed, neat and clean. Most of the farm practices are based on integrated holistic models and the herd is on track toward organics. Major health issues have been related to persistent *Haemophilus somnus* infections in 1996 and early 1997 which decreased fertility. (*Haemophilus somnus* is a gram-negative bacterium, which may live in mucous membranes of cattle. Organisms can be cultured from the nose or the reproductive tract. *H. somnus* is associated with bovine infertility and with signs of purulent cervicitis and endometritis.) Calving in this herd is year round with conventional feeding practices and highest production occurs in fall and winter. All the cattle in herd "B" are Holsteins.

Ruth
calved 1/12/96 and 3/19/97 (age 7)

checked 4/25/97—high production; soft enlarged uterus with fair tone—CLRO—mid-cycle. This cow has likely ovulated within the last 10 days.

Rx: *Calcarea phosphorica* EOD alternate with *Sepia* EOD for two weeks.

checked 7/22/97—soft uterus with mild urine pooling.

Rx: *Causticum* once daily for ten days. Milk production remains good but there is evidence of silent estrus.

Projected calving date—6/10/98.

Income over feed costs for Ruth as of January 1998 listed in excess of $5,599.

Note: Seven out of eight cows that were previously inseminated are found pregnant on this herd health exam. The majority of the herd is recovering well from Haemophilus *infection, though a few cows like Ruth did not conceive until September 1997.*

Patsie

calved 6/6/96 (age 4)

checked 8/2/96—found a large sensitive right ovary and enlarged right horn.

Rx: *Palladium* once daily for seven days.

checked 10/11/96—good uterine tone and a CL on the left ovary.

Rx: *Sepia* a.m. and p.m. for two days followed by *Ovarian* a.m. and p.m. for two days. Several cycles passed before conception.

calved 10/5/97

checked 10/14/97—evidence of cervicitis with scarring.

Rx: *Graphites* once daily for seven days.

Income over feed costs for Patsie as of April 1998 listed at minus $1,737.

(All the above symptoms are compatible with what is seen in chronic BVD-Haemophilus infection in cattle. BVD = Bovine Virus Diarrhea, also sometimes called Mucosal Disease.)

Faith

In August 1996 having calved 22 days previously (age 3 ½), she was just beginning her productive life. A low grade respiratory disease was treated with *Calcarea phosphorica*, followed by a killed BVD vaccine, and BVD nosode. Milk production is good.

pregnant—December.

calved 9/16/97—presently she is in her second lactation.

Income over feed costs for Faith as of April 1998 listed in excess of $2,278.

Homeopathic medical principles along with other holistic farming methods in a dairy herd can be a viable option. Certainly the degree of commitment for any given agricultural program plays a large role in its success. Holistic farming methods and homeopathic principles are no exceptions. Foundational to the health of any animal or herd is the nutritional status, health of the soil and the overall farm management. The medical treatment of choice should be one that enhances and does not interfere with the positive health status of the herd. Homeopathic medical principles give each herdsman that particular choice. Despite drought and other weather extremes and the presence of chronic disease, productivity and fertility can return to animals that are cared for in a holistic manner.

Making Holistic Farm Operations Work: A Case Study

Roman Stoltzfoos puts a premium on healthy livestock. A recent speaker at the Homeopathic Medical Society of the State of Pennsylvania, Stoltzfoos led the way and set the tone for the conference with a pictorial presentation of his family's organic farm—Springwood Farm.

"We prefer to let the cows do the work. Our animals are not confined; they have the opportunity to be outside, usually grazing, 365 days a year. Grazing is something any dairy manager can start immediately, with immediate benefit and very low cost. God wasn't thinking of corn when He created the cow. . . . Grass-fed is definitely the way the Creator intended cows to be fed, and we are rewarded with a superior product. We no longer grow anything on our farm that cannot be harvested by the cows." Other farm enterprises include pastured chickens, turkeys and compost. All are certified organic.

Farming for 20 years, Roman and Lucy have 10 children, all of whom are involved in the family-friendly farm environment. "We use only family labor, except for seasonal needs and family vacations. My wife and I chose to return to the farm where I was born and raised, because we saw an opportunity to work together at home. We saw an opportunity to teach our children about God's orderly creation and to teach a good work ethic."

"Homeopathy is used to treat mastitis, milk fever and emergencies. Our aim is to provide an environment and feed for our animals that will eliminate the need for most rescue treatments. The more walking, fresh air and sunshine, along with fresh green grass, that cows get, the healthier they are. This should not be surprising at all, yet it's so hard for most dairymen to change and think 'outside the box,'" Roman exclaimed. To questions about bloat, Roman responded, "When grazing alfalfa, always feed some dry hay first. If bloat should occur in the occasional animal, we give *Carbo vegetabilis* or *China officinalis* every 15 minutes. Usually, two to four doses will bring relief."

Flies are controlled with an insect sweeper. "The cattle make four passes through the unit each day, and this eliminates 70 to 80 percent of the flies. Parasites [internal] are controlled by having healthy animals and fossil shell flour."

The holding lot has been paved, and the liquid runoff is collected in a basin for irrigation. Pastures are irrigated as needed from two small ponds on the home farm. Irrigation was instituted during the dry weather of 1998 and 1999. The last year corn was planted was 1995.

"There is increased public concern about our food supply and the quality of the food available. This is to the small farmer's advantage. We Americans have been sold a bill of goods about cheap food; there is no such thing as cheap if you are concerned about quality and nutrition. In Indiana, a new dairy has been built to milk 18,000 cows. Do you want to live beside a farm that has 18,000 cows? Do you want to drink milk from a dairy that has 18,000 cows all in confinement, never getting fresh air, green grass or sunshine? Many consumers are voting with their food dollars in favor of organic food. Organic farming will save the small family farm."

"We plan to decrease the corn that is fed to zero. Economically, we will be better off. In 1998 our largest expense was purchased grain. Conventional wisdom says that you can't produce milk without grain. We believe that conventional practice is not good enough. In New Zealand, 90 percent of the farmers don't feed grain. We are no longer accepting the given method, as others do.

"In times past, if a person could not drink milk, doctors often would ask the farmer to take one cow and feed her only hay and grass (forage) and keep her milk separate from the rest. Most of the time, the sensitive person could drink that milk. Modern milk is too low in solids. It does not taste good because there is not enough (healthy) fat in it. Modern dairies skim the milk, dry the solids, then add the solids back to the skimmed milk. No wonder it doesn't taste good. Fresh raw milk from grazing animals tastes good and is good for you.

"We have found that what makes the best pasture is mix of a whole lot of different grasses. We re-seed pastures every 10 to 15 years. We cultivate or subsoil the ground at re-seeding. Every farm needs both permanent pasture and lots with fresh young grass. Cows usually go for the youn-ger grass first. With grass, a farmer gets many crops per year. You may have heard that we have to have biotechnology to feed the world. That is a false statement. We can feed the world organically regardless of what is being said in the popular press. There is too much food around of the kind that nobody wants."

Pathogens and Bacteria on the Farm

Recent scientific and lay publications have made us more aware of pathogenic bacteria that are resistant to antimicrobials. In recent years, the growth of these resistant bacteria has increased. Some scientists have referred to the phenomena as genetic drift. The microbes are mutating in response to pressure being applied to them by antibiotics. Pathogenic bacteria like *Pseudomonas* seem to be proliferating in farms across America. Regardless of the change in genetics, each resistant bacteria or organism, when it invades a body, will manifest itself in a particular way and produce a specific "symptom picture." With over 1,300 different homeopathic medicines available, and since "all of nature is our pharma-copoeia," there will be a medicine to match that symptom picture.

Our first question is how do we know that bacteria are resistant? First, clinical experiences find patients *(in vivo)* with known bacterial infections who do not recover after treatment with an antibiotic. Secondly, *in vitro* cultures (laboratory tests) indicate that there is little or no inhibition to bacterial growth on culture plates. Very often, the *in vitro* tests are in agreement with *in vivo* clinical experience.

The bacteria flora on American farms is not what it was 50 years ago. For example, milk cultures in 1950 consisted primarily of *Streptococcus,* with an occasional *Staphylococcus* or some gram-negative organism. In 1950, most of those organisms could be eliminated with penicillin. In 1999, the predominant bacteria cultured are *Staphylococci aureus,* other *Staphylococcus* organisms, coliforms, *Corynebacterium bovis, Pseudomonas* and other gram-negative non-coliform organisms. These relatively new bacteria are not eliminated with penicillin or most of the other antibacterial drugs that are available through veterinarians.

Pathogenic (disease-causing) bacteria have become more and more resistant to antibiotics since the 1950s. Both human and veterinary medicine have had to increase the recommended dosages of antibiotics in attempts to achieve the same results as were experienced with penicillin 50 years ago. The duration of antibiotic therapy today is often several months, rather than one or two weeks.

Just as herbicides destroy the weaker plants and eventually promote the growth of certain weeds, antibiotics eliminate the weaker bacteria and allow the stronger bacteria to proliferate. Within the patient there is only so much space for bacteria to live. Anti-biotics eliminate many beneficial bacteria and some pathogens, but in doing so allow other pathogenic bacteria to fill in the empty spaces. The empty spaces will either be occupied by bacteria or by the fungus and yeast families. The change in bacteria affects the entire country, and organic farms are not protected. We are culturing milk from conventional and organic dairy cattle and we are seeing similar patterns. Your neighbors who use antibiotics with their livestock have contributed to the change in bacterial population. People who take antibiotics frequently or allow a doctor to prescribe antibiotics for their family have contributed to the problem as well. For example, tetracycline has been used in human medicine for mild infections and such cosmetic conditions as acne.

To protect organic livestock, give periodic treatments of herbs and homeopathic medicines to stimulate a generalized immune response. Echinacea or goldenseal are two herbs that have been helpful in some herds. The homeopathic medicines *Sepia, Pulsatilla, Berberis, Calcarea carbonica* and *Sulphur* at proper times in the lactation cycle can give an energetic boost to the immunity. Each homeopathic medicine has a certain "symptom picture." The correct medicine is the one that has a symptom picture most closely matching the symptom picture of the individual patient or the herd.

Build up the organic matter in your fields. Feed the soil not just the plants. Plant widely diverse crops in your pastures and hay fields. Often, grasses, herbs and legumes should be seeded together in the same field. Rotate crops. It catches the pests and crop diseases off guard. Allow your livestock access to free-choice supplements like kelp and humates, while your soil is being better balanced.

Practice biosecurity. Any new animal has the potential of bringing unwanted disease onto your farm. Start to develop habits that will decrease the growth of disease-producing organisms on your farm.

Three Tips for Care of Equipment:

1. Take care of forks and shovels. Don't use the same fork to feed that you use to muck out. Keep the mucking-out and feeding forks separate. There is growing evidence that diseases such as *E. coli 0157,* Salmonella, Johne's disease and BVD are spread via feeding equipment that is contaminated with manure. Practice good management. Equipment, bunks, bowls and buckets should be scrubbed until clean from all obvious dirt and manure.

2. Rectal Sleeves. With the potentiality that BVD, leukosis and Johne's disease can be transmitted using a common rectal sleeve, it is prudent to change sleeves with each patient. Research performed in Scandinavia and at Washington State University have confirmed that this is a definite possibility. The same precautions should be observed when using needles, syringes and tattooing equipment.
3. Milking Machines. Always check the vacuum. A farmer recently related, "After changing to a totally new milking system, I was surprised how well the cows milked out." In the first week, his butterfat test went up from 3.6 to 3.9. During the previous years, it seems that most of the cows had never been completely milked out. Technology can be a harsh taskmaster. Please monitor the equipment.

Any population of animals can be referred to as an open herd or a closed herd. Open herds attempt to maintain health with rigorous vaccination programs. The immune status of each individual is frequently challenged by the entrance of new animals from many different sources and by frequent vaccinations. A closed herd or population presents a totally different situation. The immune system of each animal is less often challenged because new animals are not frequently being introduced. Ideally, all new animals are kept in quarantine for at least three weeks until it can be determined by the veterinarian on site that they are not shedding bacteria or viruses to infect the rest of the herd. Quarantine or isolation is as important for males as well as for females. In fact, often, it is the introduction of a new bull that causes the breakdown of biosecurity. If the new animals have any symptoms, test for known viruses like BVD, IBR, PI-3 and Leukosis. Maintain the quarantine until the veterinarian on site determines that viruses are no longer being shed.

All new additions to the herd should be screened for Salmonella, Paratuberculosis (Johne's disease), and *Staphylococcus aureus* mastitis as well. While it is true that cows on grass are healthier than cows in confinement, they may still be susceptible to bacterial infections. For example, *Staphylococcus aureus* is highly contagious. It can be transferred from animal to animal by the washing water, washing cloths, milking machines, milker's hands, and cow-to-cow contact. In a grazing herd, biosecurity means changing coveralls and scrubbing boots when one goes from farm to farm, and changing towels and sterilizing hands when one moves from cow to cow.

Isolate and culture all new herd additions. Milk all infected cows last. Eliminate any mud wallows in your pasture and exercise area. Make sure udders are clipped, cleaned and dried before milking. Use dry individual towels that can be rewashed or disposed of properly. Wear milker's gloves that can be dipped in boiling water for 10 seconds between each cow. Dip the milker cluster in boiling water for 10 seconds between each cow. Make sure your machines are not injuring the cow's teats and udders.

Transportation causes stress for livestock most of the time. Those who sell livestock often vaccinate with a live or modified live vaccine too close to the times of stress. The vaccine is a stress, hauling is a stress, and when combined, they can lead to shedding of virus. It has been demonstrated repeatedly that giving corticosteroids to recently vaccinated cattle causes virus shedding for seven days longer than normal. Ideally, animals should be vaccinated at least one month prior to times of transport.

Antibiotics may eliminate some types of bacteria but have no effect on viruses. Human beings have been farming since the time of Adam; it is only the last 50 or so years that agriculture has been controlled by chemicals. Most, if not all of the chemicals (that some hold so dear), will fail us. Most of the antibiotics are now failing us every day. Furthermore, research on German farms has concluded that when antibiotics are used to prevent illness in pigs, they must be dosed at high levels, resulting in the pollution of the environment. It is impossible to have healthy soil and water that is polluted with synthetic antibiotics.

The first side effect of antibiotics occurs in the gut, where beneficial bacteria are destroyed. The second side effect is a mycotoxicosis, primarily effecting the liver. While the liver is busy detoxifying, the patient is in a more vulnerable state, becoming susceptible to other "morbific stimuli." A third side effect is often seen later as an increase in chronic disease within the herd. Antibiotic residues added to water and animal wastes from confinement livestock promote the growth of toxic algae, Pfisteria, unwanted fungus and resistant pathogenic bacteria. The way to restore health is to stop using antibiotics and hormones (like BST). Also, stop applying chemical fertilizers, insecticides and herbicides to your fields. Resist the temptation to add water to manure. Instead, use plenty of high carbon source bedding. Examples are straw, wood shavings, newspaper and peanut hulls. Compost all your manure when possible. Apply the composted manure to your fields, season after season, until the organic matter in your soil exceeds six percent.

Resistant bacteria on farms is the result of genetic drift from the widespread overuse of antibiotics. The successful farm of the future must observe and practice biosecurity. Livestock need to be outside eating fresh greens daily in order to maintain health. Eating fast-growing grasses gives the greater health benefits. All purchased feeds and supplements should be certified organic; about 90 percent of the total ruminant diet should be forage. Healthy animals give us the best foods. A humane, low-stress environment for the farm family and the livestock is an attainable goal. Homeopathic treatments have been found to increase the resistance of the animal to pathogens.

A Word About Viruses

Viruses are no respecter of persons. They come knocking at the front door and find their way in, regardless of the sturdiness of the door, and whether you are farming organically, biodynamically, sustainably, and grazing. Any weakness in a system is an invitation to illness, and viruses will test you and your animals' immune competency. Fortunately, if a farm or family is prepared with a supply of homeopathic medicines, a book on homeopathic first aid, and a healthy relationship with a homeopathic practitioner, treatment and relief can be achieved at the first sign of difficulty. Conven-tional approaches to therapy involve killing the "bugs" with drugs. There are, as mentioned earlier, no commonly used drugs that will kill viruses. When respiratory viruses begin to travel through a herd of animals or a family, disease symptoms can occur. The response of each family member of any species, be it a bovine family, porcine family, or human family, is dependent upon the overall health of the immune system. And when an individual is exposed, *Aconite* should be the first defense.

Aconite and Respiratory Virus: A Case Study

One particular case where *Aconite* showed positive results involved a regular shipment of cattle from Canada to New Mexico that had routinely been afflicted with acute respiratory disease. The illness usually progressed to other illnesses, resulting in substantial losses in recent years. This year, the cattle were given *Aconite* before being loaded onto the trailer. During the trip, they drank only water medicated with *Aconite*. Upon arrival, the cattle were given a dose of Bovine Respiratory Disease Nosode. This year, every animal remained healthy.

Viruses are highly contagious. Most have an incubation period of three to 14 days. Therefore, anyone attempting to control the spread of viruses will have to isolate or quarantine all new arrivals for a minimum of 14 days. When caring for both groups, it is best to visit the sick animals after you are thoroughly finished with the healthy group. After caring for the sick group, all clothing should be changed, and boots scrubbed clean before returning to the barn where the healthy livestock are housed.

Never close buildings too tightly. Fresh air is as important as dry bedding. As soon as all the animals are well, empty the sick barn and thoroughly scrub everything down with hot detergent or steam. The barn should then be rested and allowed to dry out for at least 14 more days.

There are no more effective measures than the time-honored ones that succeed. Isolation means that sick populations are kept separate from other groups. When quarantine is enforced, animals do not leave isolation until the incubation period for the particular disease condition is passed.

For over 200 years, homeopathic medicines have been keeping people and animals well by stimulating host resistance. The mechanism of host resistance is not completely understood. When symptoms do occur, prescriptions made according to the simillimum appear to initiate a healing response. This phenomenon offers great promise for the well-being of the human and animal family.

West Nile Virus Prevention

Everyone can take steps now to help prevent West Nile Virus infections.

1. Cover or turn over all pots, cans, buckets or wheelbarrows that could catch water. If you do save rainwater, tightly cover the container between rains.
2. Remove all old tires that could hold stagnant water.
3. Clean rain gutters on buildings; make sure down spouts are clear.
4. All ponds should have fish in them to eat mosquito larvae; if fish cannot live in a pond, it should be drained.
5. Turn off all inside barn lights that attract mosquitoes and other insects at night.
6. Apply Mountain Trail Buzz-Off or *Staphysagria* to exposed skin when in mosquito infested area.
7. Avoid processed foods for people and animals at all times, and especially during the warm seasons of the year.

4.

Health from the Ground Up — Organics & Homeopathy

The workable and successful concepts of organic agricultural practices, grass-based nutritional systems and holistic livestock healthcare have enabled farm families to survive and thrive into the 21st century. At the core of any sustainable holistic healthcare system is homeopathy. It is very important for those entering organic agriculture to ponder the future and plan for any contingencies. Most of today's farmers are not old enough to remember farming without antibiotics. Because we have been accustomed to using conventional drugs, we can now substitute natural homeopathic medicines where drugs had formerly been employed.

Farmers who seek homeopathic consultation probably fall into three categories:

1. Those who are committed to holistic agriculture and will not turn back to harsh drugs.
2. Those who have tried other forms of therapy and found them wanting.

3. Those who are desperate as a result of sudden flare-ups of mastitis, high somatic cell counts, infertility or other diseases.

Homeopathy has many advantages when used in a knowledgeable manner. Farmers prefer it over harsh approaches because of its safety for the animals, farm family, and the environment, ease of oral dosing, and economic advantage. At times it may be cost-effective and preferable to view a herd of any species (especially a dairy herd) as one patient. In those cases the simillimum can be determined and the entire herd receive the homeopathic medicine.

John Muir said, "Everything is connected to everything else." Arthur Young, D.V.M., repeated that thought when he said, "A small pebble makes a large ripple in a pond. As a homeopathic veterinarian, I have a tremendous potential to have a ripple effect on the lives of both my patients and clients and on the agricultural community around me. You have this potential as well."

Some very wise homeopaths in the past saw this ripple concept and spoke about it at American Institute of Homeopathy meetings. Dr. Marion Bell Rood, a practitioner in Michigan, saw the deterioration of soil fertility and the widespread use of antibiotics eventually causing a deterioration of human health as early as the 1940s. In 1948 Dr. H.W. Eilkenberry, president of AIH, echoed the theme: "It is evident that a high percentage of our topsoil has been lost . . . yes, wasted . . . because of careless and negligent methods of farming and lumbering. Inasmuch as the topsoil is the rich and fertile part of the ground from which wholesome nourishing foods are produced, the loss of that rich and fertile topsoil has deprived us of much of the nourishment to which we are justly entitled."

The present situation in conventional and commercial agriculture emphasizes the application of N-P-K (nitrogen, phosphorus and potassium) fertilizer rather than the increasing the organic matter of the soil. "That is all the plant needs," they say. However, the truth is that the more N-P-K fertilizer is applied to the plant and the soil, the more the plant and soil become deficient. It is estimated that one pound of chemical nitrogen destroys 100 pounds of soil carbon. Levels of trace minerals such a copper, boron, selenium, zinc, cobalt and manganese continue to decline under these chemical applications. As deficiencies worsen, the rate of fertilizer application often increases until the only nutrients supporting the plant comes from fertilizer. Since the American people have been consuming these deficient foods from conventional farms (such as vegetables, meat, milk, eggs, yogurt, butter, cheese and soybeans), those deficiencies have been transferred to the public. Now, at the dawn of the 21st century, 25 percent of American couples are infertile. And severe, chronic diseases are affecting the young at much higher rates.

With organic agricultural practices, the emphasis is upon increasing the organic matter in the soil to a optimum level of four to seven percent. With the application of manures and compost derived from organic farms, the soil increases in vitality and fertility. When there is optimum organic matter, the soil acts like a sponge, soaking up the rain and preventing erosion. The much desired minerals are retained in high-carbon soil, combined in organic compounds.

With organic, grass-based agriculture, the nutrients remain in the soil, ready for plant use and in turn ready for consumption by the animals that are grazing on the nutrient-rich plant. Growth (and production) of quality vegetables and fruits is dependent on the organic matter derived from the manure and compost from organic livestock. Numerous vegetable/fruit producers testify to the need for such organic material. The decomposition of fruit and vegetable waste does not replace the nutrients that are used as those foods are grown. Animals are essential for sustainable agriculture.

TWO ORGANIC FARMS

FARMER A

Purchased farm 1992
Began dairying 1995
First homeopathic case . . . . bloating herd
Herd size 34
Neighbors . . . . 12 cooperative
Work force . . . . parents do 99%
Grass based . . . . semi-seasonal
Certified 1997
Sells . . . . milk, beef, farm calves

Regular reproductive exams . . . . since 1997
Herd consists of . . . . . Jersey, Holstein, Brown Swiss, Ayrshire
First therapy . . . . Homeopathy
Meds used first year . . . . Sepia—missed estrus
Ovarian—regulate ovulation
Carbo—veg bloating, alt. with *Nux vomica*
Frequently used medicines at present . . . . . *Carbo veg, Arnica,*
Phytolacca, Aconite, Lycopodium, Pulsatilla, Sepia
Recent somatic cell count 160,000
Goals . . . . "To produce the cleanest, freshest and most
nutritious milk possible, and to use nature
as our main partner."

FARMER S

Rented farm 1985-1996
Purchased farm 1997
Began dairying 1985
First homeopathic case . . . . paralyzed goat
Herd size . . . . 100+
Neighbors . . . . 5 skeptical
Work force . . . . parents and nine children do 99%
Grass based . . . . semi-seasonal
Certified 1995
Sells . . . . milk, cheese, beef, eggs, chicken,
turkeys, pigs, replacement heifers
and organic compost
Regular reproductive exams . . . . since 1991
Herd consists of . . . . Holstein, Dutch belted, Jersey,
other crossbreeds
First therapy . . . . Homeopathy plus some herbs
Meds used first year . . . . *Calc carb, Calc phos, Pulsatilla,*
Ovarian—all four meds used to regulate estrus.
An acute uterine torsion case responded to
Caulophyllum, Arnica and *Pulsatilla.* Following
delivery, uterus was flushed with *Calendula,*
Colostrum and sterile saline.
Frequently used medicines at present . . . . *Apis malefic,*
Pulsatilla, Ovarian, Cimicfuga, Sepia, Murex, Calc carb,

Silicea, PTT (Pituitary Posterior, Thymus and Thyroid),
and Mastoblast H.P.

Recent somatic cell count 210,000

Goals . . . "To have a farm where our children would be able to work with their parents and to farm in such a way to make it attractive to the next generation."

These two organic farms in Pennsylvania have unique and different histories. Farmer A bought his farm in 1992 while a member of the state government. At that time, he and his family made a commitment not to use any chemical fertilizers. They planned that when his time in government service ended, the farm could be developed into an organic dairy. In 1995, Farmer A began his grass-based dairy operation. An emergency situation of bloat occurred. A third of the herd had symptoms of bloat. The local veterinarian, though apologetic, offered no non-conventional solutions. The farmer quickly sought homeopathic consultation. With frequent dosing of *Carbo veg,* all the cows responded and not one was lost. That tense time was the springboard that confirmed his need for and belief in homeopathic medicine for his 34-cow herd.

Farmer S began conventional farming in 1985, but after three years his hard work matched his frustration. Initially very skeptical, Farmer S would sit back in the corner during my homeopathic lectures and asked questions that put me on the spot. He just wasn't convinced of the ability of homeopathic medicine to address the many health problems in his livestock, but he was an honest seeker of truth. Then a family pet goat became paralyzed. With no conventional drug options and pressure from his children, he relented and called for homeopathic consultation. *Cicuta virosa* was the remedy chosen, with positive results. As the goat im-proved, the medicine disappeared, only to be found in the hands of their two-year-old, the bottle empty. With assurances from their homeopathic medical physician that no harm had occurred, they wondered, "What other medicine could bring a paralyzed goat to its feet but not harm a two-year-old child? Only a homeopathic medicine." That incident was the springboard that began their homeopathic adventure for their 100-plus dairy herd, and other species of livestock on the farm.

Farmer A's farm, although mildly neglected, had not been abused by chemicals before he purchased it. It is often easier to take a neglected farm and return it to top condition than to take a chemically abused farm and try the same thing. Because an organic farm needs a 25-foot buffer zone surrounding it, a wise and courageous farmer will seek the support of his surrounding neighbors. This he did. All 12 neighbors agreed not to apply any forbidden fertilizers or chemicals on their 25 feet of property that joins Farmer A's land. That in itself was quite an achievement. It then was relatively easy to have every square inch of his land certified, followed by the dairy cattle. Farmer A's farm would best be described as a semi-seasonal, grass-based organic dairy. The missing link to his operation was, of course, homeopathy.

Farmer S is surrounded by five other farmers, all skeptics. Even though they thought and communicated to him that he was "cracked," he continued to change methods toward sustainability. The 25-foot buffer zone is maintained by Farmer S on his own land. At present, he and his wife and ten children operate a 280-acre grass-based semi-seasonal organic dairy.

Many of the universities find it difficult to accept the fact that an organic dairy operation can be profitable without having the farm family rely on "off-the-farm income." All the income of Farmer S is derived from the land, the crops, and the livestock.

Kinds of Dairying

A grass-based dairy operation means:
1. All of the farm produces grass or legumes for the purpose of grazing or hay making;
2. Only small amounts of grain are fed to the animals, or none at all; and
3. The farm has only a few acres of non-grazing crop land compared to the large percentage of the land that can be grazed. "If you can't graze it, you don't raise it."

A semi-seasonal dairy operation means:
1. The majority but not all of the livestock have their calves the same time each year and therefore are dry or non-lactating at the same time as well; and

2. In Pennsylvania, births occur in March and April and the
   cattle are dry or non-lactating in December, January
   and February.

A seasonal dairy operation means:

All the cows would be giving birth within a four to six week
window of time and be dry or non-lactating nine months later
when they and hopefully the farmer and his family go on vacation
(or attend a homeopathic meeting).

A conventional year-round dairy means:

Constant calving at all times of the year, leaving little room for
time off for the farmer and his family.

Forage, the Preferred Feed

God did not intend cows to eat grain. Cows do not need to eat
soybeans. The cow was not created to eat the things man eats.
Ruminants convert forage into milk and meat and hide and things
for us to use. A conventional farm is feeding grain to these cows
by the shovelfuls and then they have to give drugs in large quantities to combat the acidosis produced by the high levels of grain
and the stress of confinement and crowding. By contrast, organic
livestock in a grazing system have little indigestion and live a low-
stress lifestyle.

Chemical fertilizers were first promoted in Europe and then in
North America in the early 20th century. By 1950, chemical fertilizers had replaced composted manures as the most frequently
applied soil amendment. Many of the world's farmers had become
convinced that all you need to do for larger yields was to put a
little-N-P-K fertilizer on the plants. The nitrogen, phosphorus,
and potash compounds on the plants would produce faster growth
as well. The farmers of the early 21st century are paying for the
sins of the farmers of the 1940s and 1950s. Fortunately, Farmer
A and Farmer S are not going that route. As Farmer S would say,
"I'm religiously opposed to chemical fertilizers."

In 1990, our practice serviced one organic dairy in Vermont.
Opportunities developed to allow my wife and I to lecture and
instruct farmers and veterinarians in the principles of homeopathy
for the health needs of their organic farms. At present there is a
dramatic increase in the number of organic dairies. In the first

nine months of 2000, about 1,300 dairy farms were certified in the United States. It is growing at about the same rate as homeopathy—25 to 35 percent growth per year. Now the consumer is no longer limited to specialty stores but can purchase their organic vegetables, chicken, turkey, yogurt, cheese, milk, eggs and meats directly from the farmer.

Farmer A, Homeopathic Medicines

What homeopathic medicines did Farmer A use his first year? For the bloating symptoms, *Carbo veg* was dosed frequently. Later, when a new pasture was opened, the drinking water was medicated with *Carbo veg* and cows were encouraged to eat a little dry hay and take a drink before grazing. A few stubborn cases of bloat were dosed with *Nux vomica* in alteration with *Carbo veg*.

The first year of farming *Sepia* was prescribed for each missed estrus. A dairyman knows that it is too late to breed a cow when he observes blood-tinged mucus on her tail. Ovulation has passed. After *Sepia* she will again be receptive in 19 or 20 days. Cows in proestrus were given *Ovarian* before each breeding. This homeopathic nosode prepared from the fluid of a healthy ovary helps to regulate ovulation.

Today, Farmer A employs *Arnica* in cases of trauma, *Phytolacca* in painful mastitis, and *Aconite* for acute fevers. The next most frequently used medication is *Lycopodium*, which is effective in the prevention and treatment of the metabolic condition known as ketosis.

Farmer S, Homeopathic Medicines

In his first homeopathic year Farmer S found that *Calcarea carb* and *Calcarea phos* were strongly therapeutic in maintaining milk production and fertility in the herd. Calcium was likely deficient over the entire farm in those early years. Conventional farming practices and N-P-K fertilizer often produce deficiencies of calcium, carbon, and trace minerals. Applying manure and compost year after year will replenish these soil nutrients.

In addition to *Sepia* in post-estrus or following ovulation, dosing with *Pulsatilla* in proestrus and *Ovarian* in estrus was helpful for the herd during the 1991 breeding season. Regular herd health exams continued for two years with no major episodes. In 1993,

Farmer S experienced a rash of illness in livestock. The cause was found to be mold in the corn silage. Afterward, the family began diligently seeking a feeding program that did not rely on corn silage.

Both of these family farms have enjoyed some measure of economic freedom since converting to grass-based organic dairying. Farmer S in 1998 recorded an income of $764 per cow per year. His cull rate was 18 percent. The national conventional average is 40 to 50 percent, and the organic cull rate average is 30 to 33 percent. These two farm families are examples of success in organic dairying. Using homeopathic medicines (and principles) lead to both success and sustainability. If questioned, I am sure that each family member would be enthusiastic about the progress of the past and are making plans for more sustainability in the future.

Intensive Grazing Practices

More Examples from On the Farm

Farmer A began farming in 1991, started using homeopathy in 1994 and initiated intensive grazing management in 1995. His careful rotation schedule and judicious use of the back fence has enabled his adult cattle to remain nearly parasite free. The entire herd, including weanings, yearlings and bred heifers, are treated spring and fall with homeopathically medicated water five days in a row. The homeopathic medications used are *Santoninum, Chenopodium, Granatum* and *Abrotanum*.

Farmer A has a very colorful herd of dairy cattle including Brown Swiss, Jersey, Milking Shorthorn, Holstein and crosses of each. Monthly herd health exams from March to December are followed by dosing with the prescribed medication from his homeopathic first-aid farm kit. Each cow is internally examined when necessary, and a specific or constitutional medicine is prescribed. Most frequent prescriptions are *Sepia, Phytolacca, Pulsatilla, Graphites, Phosphorus* and *Silicea*. Occasionally, there is a need for *Calc phos, Calc carb, Natrum mur* and *Ovarian*.

When the somatic cell count (SCC) began to rise above 250,000 in 1998, milk samples were taken and cultured. A herd nosode was prepared from the milk and the cultures. Dosing the high cows with *Calc phos* and the nosode or *Phytolacca* and the nosode lowered the SCC to 220,000 in one month. Percent successful services from 3/98 to 1/99 averaged 71 percent. Lancaster Company DHIA averaged 37 percent for that same period.

As soil fertility, cow health and pasture quality simultaneously improves, Farmer A is reaping health and financial benefits for his family. By 1997, Farmer A had finished the transition period to become certified organic, and the farm was positioned for future success. Remember, he has accomplished this health level without the use of antibiotics, hormones (GRNH, BST, etc.), dry cow treatments or chemical wormers.

Farmer B has been using homeopathic medicines on his livestock for three and a half years. He and his family milk 40 to 44 purebred Holsteins year round. Using a balanced approach in a transition system, he has changed his treatments of the soil, the crops and the animals simultaneously. He was one of the first farmers in his county to use energy medicines to inhibit weeds in his pasture.

Percent successful services for the last 12 months averaged 54 percent. Both herds use a tie-stall barn, Intensive Grazing Management, free choice minerals, and both use a bull and artificial insemination (AI). Farmer A grazes year-round, while Farmer B grazes about ten months of the year.

Farmer A chose homeopathy to improve milk quality and conception rates, but Farmer B changed on the recommendation of his milk inspector. For whatever reasons the change was made, the transition has proven beneficial for the land, the animals and the farm families. Many farmers who become certified organic grazers find that herd health problems decrease in frequency. Free choice minerals are offered year round to balance the livestock as the land and soil are gradually being re-mineralized. Occasionally other nutritional supplements are offered to boost immunity during times of stress.

Controlling Somatic Cells

This is the story of an organic farmer in New York. He was certified organic in January 2000 after five years on his farm. On his previous farm in Pennsylvania, his SCC had always been below 200,000. For the first years in New York, the count was almost always above 400,000. Several specialists examined the cows, the milk, and the farm, but could not explain why the SCC remained so high. They had no recommendations to remedy the situation. Since the farmer and his family were not given any good reason not to become certified, they went forward and completed the process. In 1999, they could often smell natural gas in the drinking water, which contained *E. coli* bacteria on most tests. Thirty percent peroxide was being metered into the water to decrease the bacteria.

To help reduce the SCC, Mastoblast (original formula) was employed to treat all lactating cows for ten days during early January 2000. By February, the herd SCC was already decreasing. Milk cultures from the five highest cows were plated out by the Clark Veterinary Clinic for bacterial identification. Culture plates produced the following bacteria: *Staph aureus, Staph epidermis, Streptococcus uberis,* and hemolytic *E. coli.* These were homeopathically prepared according to HPUS and stored for nosode preparation. (See chapter on *Materia Medica* for more information on nosodes.) The farmer stated that Mastoblast was given to the dry cows in March. In April, the lactating herd received Mastoblast HP. Herd SCC through the summer of 2000 stayed below 160,000. In September, the herd SCC was 330,00. This may have been a result of the well water not being completely free of bacteria.

"One cow, which we raised as a calf, has been on our farm for over six years. For all four lactations, her SCC always ran between 750,000 and 1,000,000, but, her annual milk production was between 25,000 and 30,000 pounds, so I didn't feel I could sell her. Mastoblast and *nosodes* brought her count down to 50,000 at the last drying off. She recently freshened (calved) and her count was still low and she has had no mastitis."

"In February, we had three sets of twins. That scared me. Cows tend not to clean after having twins, and they are difficult to breed back. Utilizing homeopathics, two bred back in 65 days, and one took 80 days."

"Animals need space. You can buy expensive heifers, but if they are overcrowded, that will increase their stress and cause them to be stunted. Take better care of what you have, and you'll do better. We use the pastures extensively (and intensively). There are 21 paddocks and each one is 1.6 acres. We rotate to a new pasture every 24 hours. At the end of the grazing season (November) we will apply our compost and amendments. We apply nothing during the summer.

"Dry cows are fed whole oats and silage and Fertrel minerals. If there is any mastitis, we believe in frequent stripping and dosing with the homeopathic medicines listed in the homeopathic farm kit insert booklet."

As evidenced by these on-farm examples, there are many factors—space for animals, cleanliness of water supply, nutrition, herd history, previous treatment with chemicals and vaccines, condition of grazing land, etc.—that all contribute to the health of the herd. The point is that by utilizing holistic and organic farming practices along with using homeopathic medicines first, a farmer can go a long way toward reducing on-farm problems and disease.

5.

Prescribing
in Homeopathy

In homeopathic medicine, keynotes refer to prominent symptoms that the patient will show when a particular homeopathic medicine is indicated. Like a jigsaw puzzle, each homeopathic medicine is the individual piece that only matches that empty space within the picture. The empty space in the picture represents the symptoms of the patient. Matching the puzzle piece to the space results in proper prescribing. It sounds difficult, but with time and practice it will become second nature.

Modalities are environmental "conditions" which make the patient better or worse; they modify the symptoms. Pellets are the pill form of medicines and dilutions are medicines in 20 percent ethanol and 80 percent distilled water. Keynote symptoms are the ones most helpful when confronted with specific disease conditions in livestock. The conventional approach concentrates on the organism—bacterial, viral or protozoa, whereas, the homeopathic approach focuses on the patient. The patient has both general and particular symptoms. General symptoms affect the entire patient, while particulars affect a part of the patient. An example of general symptom is lameness. Leading to the lameness might be a fractured bone which is a particular symptom. Both are equally important and must be included as keynote symptoms.

When being confronted with diarrhea in one or more animals, find the heading of diarrhea in the Index or Specific Conditions chapter of this book. Next, while observing the animals for more symptoms, ask questions similar to the following. What are the consistent symptoms that a majority of the sick animals are showing? Are they weak in pain? Do they appear chilly? Are they feverish and panting? Are they thirsty? How often do they drink? What environmental conditions make them worse? What changes make them better?

Success in homeopathic prescribing is greatly improved when one can match three or four prominent diseased symptoms of the animal with three or four keynotes of the homeopathic medicine. The goal is always to match as closely as possible symptom to keynote.

Dr. Samuel Hahnemann recorded his exhaustive studies in his treatise, *Organon of the Medical Art,* which has been superbly edited and annotated by Wendy Brewster O'Reilly, Ph.D. The following list comes from her 1996 publication of *The Organon.*

The most appropriate medicine for a given case of disease is the one which is the most homeopathically selected and which is administered in the correct dose.

The more homeopathic a remedy is to the disease, the more curative it will be when given in appropriately small doses which act gently.

The best guides to determining the appropriate dose in each case are pure experiment, accurate experience, and careful observation of each individual patient.

As a rule, even the smallest homeopathic dose will be strong enough to begin a cure, except in cases in which vital organs are already seriously damaged, or in which other medicinal substances interfere with treatment.

Medicines may be administered by olfaction (gentle inhaling).

Administration of medicines may be though the breast milk. (Give mom the remedy when she is nursing the babe.)

The medicinal solution may be rubbed in at the same time that it is ingested.

Many homeopathic medicines will function in cooperation with each other in order to complete the healing process. One could liken it to a relay race; each time the baton is passed to a different runner until the race is completed. All homeopathic medicines should be stored away from heat, bright light, strong odors, and from electromagnetic fields such as motors. Never store them near microwaves or use a microwave to heat a homeopathic treatment of any kind. It is a good idea for owners to keep their farm homeopathic medicine kit in a previously designated location in the barn or milk room.

Acute vs. Chronic Illness

When a veterinarian or farmer is faced with an ill animal, it is advantageous to determine if the symptoms are truly acute. Do they represent a self-limiting condition which will soon resolve itself with the proper prescription? Or do those symptoms represent an illness that will require a second prescription to bring the animal back to a healthy state?

Acute conditions with clear, uncomplicated symptom pictures call for acute medications. For example, *Aconite napellus* is useful in disease states characterized by fear along with a sudden increase in body temperature. *Arnica montana* is indicated for soft tissue trauma with bruising and bleeding. *Hypericum perforatum* is indicated in injuries where there is nerve pain with numbness or paralysis.

Functional change means decreased ability. A cow with a teat fibroma has decreased ability secreting milk past the obstruction. A horse with a torn ligament has decreased ability to work or perform. Functional changes respond well to the proper homeopathic medications, but waiting weeks or months before treating will decrease one's success. Improper treatments or lack of any therapy may lead to a chronic condition. Some treatments can actually produce more chronic disease due to suppression. Inasmuch as it depends on you, never be the instrument of suppression.

The Chronic State

All animals are born with a degree of subclinical chronic disease. With age and suppressive drug use chronic disease will

CHRONIC DISEASE

Symptoms are visible in this region

Crises

Time

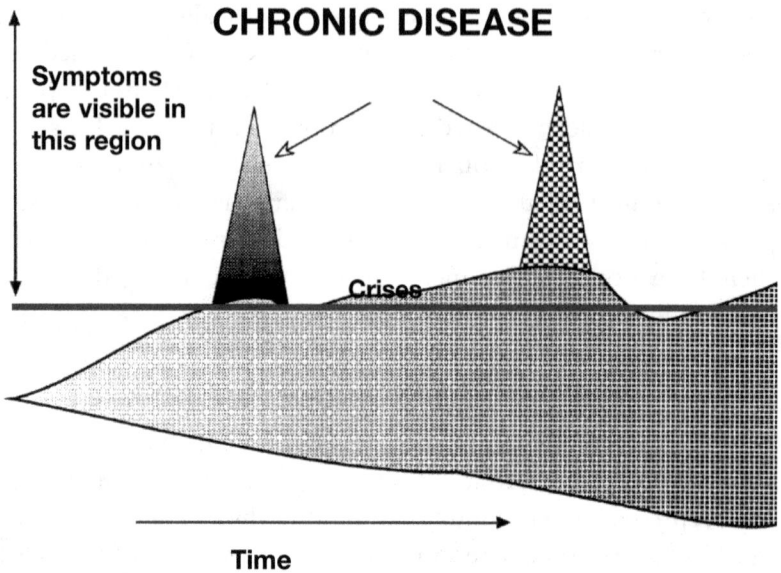

Chronic disease exists primarily "below the surface," expressing itself as continued symptoms that, at first, do not seem severe and punctuated with periodic explosive crises. These crises are often given different diagnostic names but are actually the same disease taking different expression as the disease evolves or is modified by treatment. Illustration by Richard Pitcairn, DVM, used by permission.

increase and be manifested. The greater the use of suppressive drugs, the greater the suppression of the immune system thereby causing an increased level of chronic disease in the animal. Suppressed patients have organ and tissue pathology which is more difficult to heal. These cases require more work and more time. The patients are more likely to experience an aggravation or a healing crisis.

Note: *Aggravation* is a state where symptoms are temporarily worse while on the prescribed medication. Many homeopathic physicians like to see aggravations in their patients because it indicates that the medicine is the correct one. In a true homeopathic aggravation, even though the symptoms are worse, the patient says, "I feel better." A *healing crisis* is a point in the course of an illness when illness can only go one of two directions. Often this is experienced in influenza when a person slowly develops symptoms which worsen to reach an apex. Then the fever breaks and the patient rather rapidly begins to recover. In veterinary prescribing, aggravations are more common than healing crises. In animals, aggravations last an hour or two, but may occasionally go on for a day.

In chickens, a broiler may not live long enough to manifest significant chronic disease, whereas, a laying hen has more opportunity to do so. A veal calf would have less subclinical disease or hidden chronic disease compared to a two-year-old steer raised in a conventional feedlot environment. Breeding animals, because of age, reproductive stress and general suppressive drug use, will need homeopathic treatment to counteract the degree of chronic disease they are carrying.

The occurrence of internal chronic disease may be likened to an iceberg (see figure above). Only a small part of an iceberg is visible above the surface of the water whereas the larger portion remains hidden under the water. So too, is the chronic disease state. More is under the surface than above the surface. As each animal experiences an illness, if the animal is treated with suppressive drugs, the chronic disease below the surface increases. Above the surface the animal may appear well but the residential remains within. Acute disease, on the other hand, is readily observed by all. Only a small percent of the total chronic disease is visible to the casual observer.

A dairy farmer may be pleased that his herd is considered healthy due to high production, but he/she has not been trained to observe below the surface. It is only when there is an acute flare-up of symptoms from stress or other "morbific stimuli" that there is an indication that all is not well. (Dr. Samuel Hahnemann defined "morbific stimuli" as that which causes symptoms to arise and hence causes disease to manifest itself.)

Certain treatments add to the degree of chronic disease or the volume of the submerged iceberg. Unnecessary multivalent vaccines have been shown to cause illness in dogs, cats, horses, cattle and other species. Many illnesses become chronic when treated with anti-inflammatory drugs. These drugs may reduce suffering temporarily, but do nothing to augment the healing process. When the immune system is suppressed, more chronic disease results.

How many generations of dairy cows, for example, have been treated with corticosteroids for symptoms of ketosis? We know that corticosteroids are suppressive medications. How can we expect to have a healthy immune response in a cow who has been treated with suppressive agents, whose dam had been treated with suppressive agents before her, and whose grandam had been treated with suppressive agents before her? Fortunately, once a farmer feeds a natural organically grown ration which utilizes forages over concentrates, the condition called ketosis (acetonemia) becomes only a memory and a condition of the past.

What measures can be taken to reduce chronic disease in an individual or in a herd? Feed organic. Vaccinate only for those diseases in your neighborhood. Follow vaccines with oral nosodes. Use antibiotics sparingly or not at all. Use homeopathic medicines early and complement them with probiotics. Take good care of your fields, especially your grasses. Use Primary Spray Conditioner on manure and compost to help hold nitrogen and other nutrients.

The First Medicine

When acute symptoms do occur, the simillimum is your best guide. What four symptoms are most important? Which most closely match the symptoms of the medicines that you have in your homeopathic first aid kit? Many acute diseases will be cured with one medicine. When you are dealing with an acute manifestation of a chronic disease, two or more medicines may be needed. For example, *Belladonna* or *Phytolacca* often cure acute mastitis. When the somatic cell count remains high even though no clinical mastitis is present, a second medication is in order. A homeopathic prescriber must rule out all other physical or mechanical causes and then retake the case. Setting aside the original symptoms of the patient, he bases the second prescription upon the four or more symptoms that he sees at the present time. If yesterday's symptoms were included, it would blur today's symptom picture.

The Second Medicine

Medicines that follow *Belladonna* well are *Calcarea carb, Hepar sulph, Lachesis, Phosphorous, Pulsatilla,* and *Sulphur.* Medicines that may be given after *Phytolacca* are *Arsenicum album, Bryonia, Calcarea carb, Hepar sulph, Lac caninum, Lachesis,* and *Rhus tox.* The second prescription will often remove some of the chronic dis-ease present in the patient (or in the herd).

A word of caution. Unless you have been practicing homeopathic medicine in excess of 40 years, do not try to shortcut the healing process of the body. Give the acute medicine first; wait for the response to finish, and then give the second medicine.

My client Mr. G. has Holstein dairy cows. Last winter, there were several with acute mastitis and high fevers. Udders were hot and so was the milk. The cows were highly excitable, had dilated pupils, and dry mucous membranes. *Belladonna,* the prescription of choice for the acute picture, cured all symptoms except for hardness of the tissues and occasional flakes in the foremilk. Other chronic symptoms included being easily chilled by cold, wet weather and a SCC above 400,000. *Calcarea carb* best matched the symptom picture for the second prescription and brought the cows back to health.

Acute Becomes Chronic: A Case Study

An abscess occurred in a horse several days after a vaccination. Generally, abscesses treated homeopathically utilize three particular medicines in specific order. The first medication should be *Hepar sulph.* When the abscess is completely drained, follow with a few doses of *Silicea* to aid healing from the inside out. A few patients will require a third medicine, *Myristica,* to complete drainage and healing. One might say that 75 percent will respond to *Hepar sulph,* 20 percent of the remaining will respond to *Silicea,* and the final 5 percent will respond to *Myristica.*

In this case, the owner elected to give *Myristica* first. No response occurred. Later *Silicea* was tried with only a weak response. An acute case that should have resolved in three to five days dragged on to become a chronic one. A severe illness resulted with cellulitis, lymphangitis and diffuse infection. For several days the fever was 104 degrees, the appetite depressed, and the animal could move only very slowly due to swellings of the hip, leg, flank and chest. Intensive supportive therapy was required

---

including daily soaks, probiotics and, reluctantly, antibiotics. After four weeks, the abscess burst and drained. Using careful prescribing of homeopathic medicines, the equine patient was slowly guided back to health. Convalescent medicines were *Sulphur* 30C daily for five days followed by *Silicea* 200C once weekly for three weeks, then *Thuja* 200C once weekly for three weeks.

All animals have a measure of chronic disease within. If not properly prescribed, both conventional and homeopathic drugs can add to the level of chronic disease. The human body and the animal body heals by known physiological and biochemical mechanisms. One needs to choose a medical regimen which will augment rather than hinder that process.

The goal of every person in agriculture should be healthy livestock, healthy soil, healthy plants and healthier people. Mature breeding livestock carry subclinical disease, which interferes with good health and production. Our goal as homeopathic prescribers is to provide quality care which will not add to the burden of subclinical disease. When we lose sight of our goal and the future becomes foggy, we can get ourselves into trouble.

The uneducated person may not be aware of the presence of chronic disease. The farmer knows that he/she is losing money. They may have been told by specialists that a high cull rate or high death rate is normal. One farmer with a 300-cow herd reported that he had six cows die in one week. He was told by his veterinarian on site that it was "nothing to be concerned about." Surely subtle signs of stress or illness were apparent long before six deaths occurred.

Going back to the use of synthetic pharmaceutical treatments will lead to grave consequences for the future. In very difficult times a farmer can find himself looking back rather than forward. A herd that has been accustomed to natural medicines like *Pulsatilla* and *Sepia* will quickly suffer if again placed on hormones and antibiotics. Under the influence of these suppressive drugs, both clinical and subclinical disease will increase. Instead of each animal's immune system being alert and ready for any challenge, it will be subdued and sluggish. A new homeopathic prescription will be needed for a wake-up call.

Chronic disease tends to be invisible to the casual observer. Only when an acute crisis occurs do we realize how much chronic disease may be present below the surface. If the patient is again treated with suppressive drugs (antibiotics, steroids, etc.) the burden of disease becomes greater.

As an example, the first two times a patient had an acute crisis, he was treated suppressively, causing chronic disease to increase. The third time an acute crisis occurred, perhaps the care giver had learned that previous approaches were faulty. This time, a complementary therapy was chosen. A complementary therapy works hand in hand with the immune system, complementing its action and thereby removing the acute disease, as well as some chronic disease. The fourth crisis was again handled as the third had been, using a complementary treatment rather than a suppressive one. The patient recovers more quickly, and the burden of chronic disease becomes lighter. The diligent homeopathic prescriber will observe the improved energy and vitality in the patient. Improvement can be expected to last for not just a few days, but for months and years.

Miasm

Dr. Samuel Hahnemann, through many years of observation and research, had come to the conclusion that there are deep predispositions to disease in each and every being. These deep predispositions he described as miasms. Dr. Day, in his book *The Homoeopathic Treatment of Beef and Dairy Cattle,* gives the definition of miasm as "a Hahnemannian philosophical term for an infective agent." No direct definition exists in modern medical terminology. The three miasms Hahnemann described were Psora, Sycosis and Syphilis. The first two, Psora and Sycosis, have influence in veterinary homeopathy. The underlying miasm called Psora relates to hereditary disease passed down from one generation to the next. Every person and every animal has a measure of Psora, and some have more than others.

The second of Hahnemann's miasms he called Sycosis. Causes of this underlying state could include unhealthy coition and venereal disease, bites and stings from cats, dogs, rats, insects and poisonous snakes, and injections (vaccinations) which penetrate the integument of the body. When the body's integument is penetrated, there is always a risk of compromising the integrity of the immune system.

Uncovering Vital Force

Chronic disease is often likened to layers or blankets over the patient's vital force or defense mechanism. Because of the oppression of miasms, the patient is hindered in his expression of life and health. The lower blanket is likely to be Psoric, with a specific symptom picture. The next several blankets will be Sycotic, each also having a specific symptom picture. The top blanket will be the present acute condition.

The most expedient approach removes the top blanket first by prescribing a medicine that best fits the current disease picture. The time involved in removing a disease layer varies greatly. A *Belladonna* case may resolve in a matter of hours whereas a *Thuja* case may take several weeks to clear so that the next layer can be clearly identified. The veterinarian, the patient and the owner all must exercise patience to see the process through to completion. Do not prescribe for the next layer until it can be clearly discerned.

Picture someone trying to move with ten blankets on top of their body. There is little freedom of expression. It is difficult to discern the shape or the vitality of the body under the blankets. Perhaps after half of the ten blankets are removed, one is able to see a live, breathing, moving person. The true homeopathic doctor will continue to work with his patient to enable him/her to become as free as possible.

Patients with skin and joint manifestations have more visible symptomatology than patients with chronic disease involving vital internal organs. The invisible portion of the iceberg becomes larger when internal organs are involved (see illustration).

The goal of proper therapy is to systematically remove each layer in succession in order to uncover the core of the patient's vital force (defense mechanism). As each blanket is removed, the patient is enabled to express himself more freely.

Out of the Past: A Case History

The following case history, which took place in the late 1800s, was presented by James Tyler Kent, an M.D. from St. Louis. The case illustrates several truths that are observed in patients suffering with chronic disease. Because human beings are inherently more complicated than livestock, and because of increasing longevity and other factors, chronic disease plays a greater and greater part in our health and well-being. By studying human cases, we can learn a great deal about chronic prescribing for animals.

At 38 years of age, Mrs. N had been an invalid for about ten years as a result of chronic arthritis of the left knee. Two well-respected surgeons had not been able to arrest the previous acute disease present and had recommended amputation of Mrs. N's leg. Dr. Kent's care utilized five different medications throughout the treatment period. On several occasions, this patient did experience an aggravation for several days after each medicine was given. More importantly, both Dr. Kent and the patient kept working toward the goal of improved health. Dr. Kent not only illustrates superb technique in chronic prescribing, he also offers the patient's perspective throughout the 30-month treatment. For a livestock owner, the goal is improved health for animals, soil, plants and his/her family. For the homeopath the goal is always to heal the patient and to cure disease.

July 16, 1881: Patient in bed; entire limb wrapped; great pain on motion; great burning in soles of feet. Rx: *Sulphur* 55M, one dose. For several weeks, patient suffered greatly at night.

August 20, 1881: Beginning to improve; patient says she is greatly relieved after the aggravation has passed. Rx: *Sulphur* 81M. A few days of pain, then much relief.

September 20, 1881: Mrs. N's husband was severely ill, and she cared for him day and night until he died. (It is amazing that already she had improved enough to care for her dying husband). Rx: *Ignatia* daily until grief has passed.

October 8, 1881: Repeat of Rx: *Sulph* 81M.

November 12, 1881: Joint definitely smaller, foot no longer oedematous or burning; Appetite good, gaining strength. Wait. [Note: The term *Wait* means that the doctor believed that the last prescription continued to act. He chose not to interrupt the action and improvement by dosing until the patient's body actually called for the next dose of medicine.]

December 3, 1881: Cold weather, feet and leg cold, pain in knee and craving for eggs. Could not become warm. Rx: *Calc carb* 85M.

January 7, 1882: More mobile in the house, even though flexion causes great pain. Mrs. N is able to get out of bed holding the leg up (with the other foot) and then can move about the home comfortably on crutches. Wait.

February 3,1882: Gradual improvement. Repeat a dose of *Calc carb* 85M.

March 25, 1882: For the first time in ten years, Mrs. N can now wear a normal size shoe on both feet. Wait.

April 4, 1882: Improved and at a plateau. Rx: Repeat *Calc carb* 85M.

May 30, 1882: No new symptoms; eating well; sleeping well. Wait.

June 3, 1882: Sour eructations; restless nights; knee more painful. Rx: *Lycopodium* 71M. Afterwards, knee so painful that patient was compelled to stay in bed several days. Dr. Kent visited her daily and helped with her care.

July 2, 1882: Walking on crutches with very little knee pain. No more stomach symptoms. Wait.

August 3, 1882: Improving. Wait

September 2, 1882: Plateau. Rx: *Lyco* 71M.

September 6, 1882: Slight aggravation and then continued improvement.

October 1, 1882: Improving. Wait

November 8, 1882: Improving. Wait.

December 15, 1882: Repeated a dose of *Lyco* 71M.

January 1883: Mrs. N continues to improve. Dr. Kent supplied her with a cane to use instead of crutches. She no longer fears knee being bumped. Wait.

May 1, 1883: Mrs. N can now bear some weight on left foot. (First time in 12 yrs). Rx: *Lyco* 71M, one dose.

July 8, 1883: Rheumatic pains both knees. Restless at night, better during continuous motion. Rx: *Rhus tox* 1M, doses every three hours for two days.

August 5, 1883: Improving. Rx: *Rhus tox* 32M, one dose.

September 1, 1883: Walking all over the house with one cane. Happy. Wait.

October 1, 1883: Improving. Rx: *Rhus tox* 32M, one dose.

November 2, 1883: Rx: *Rhus tox* 32M, one dose.

December 1, 1883: Patient walked to Dr. Kent's office with aid of cane and walked in without it.

January 7, 1884: Patient again walked to office without cane and has a limp but no pain. Dr. Kent asked if she regretted two and one-half years of homeopathic treatment, to which she answered, "Ten thousand times, no."

Constitutional Prescribing

Acute and chronic prescribing should improve the patient's overall health and at the same time not add to the burden of chronic disease. As the earlier illustration showed, chronic disease is as a submerged iceberg and, like an iceberg, most of the mass is beneath the visible surface. The goal of constitutional prescribing is the removal of hidden invisible chronic disease. Christopher Day, MRCVS, defines a constitutional remedy as "one which takes the entire makeup of the patient into account, rather than presenting symptoms and concomitants alone. It matches the pattern of the individual body's programmed response to disease. This type of remedy has a significant effect on every organ system, including the mind." He defines a polycrest as "one of the deep-acting, extensively applicable remedies which have a wide action on all parts of the body. Constitutional remedies are polycrests."

"The constitutional drug is that which is most effective in treating the individual in the sum total of his strengths and weaknesses: mentally, emotionally and physically," said the late Margery G. Blackie, M.D. in her book *The Patient, Not the Cure.*

One could liken a constitutional prescription to a gentle spring rain—nourishing the environment thoroughly. In the text *Hahnemann's Chronic Diseases,* Dr. Hahnemann emphasizes the importance of giving the "constitutional" or "antipsoric remedy" during pregnancy. He wrote, "Pregnancy in all stages offers so little obstruction to the antipsoric treatment. . . . Most necessary, because chronic ailments are more developed during pregnancy. In this state of women, which is quite a natural one, the symptoms of the internal psora [submerged iceberg] are often manifested most plainly on account of the increased sensitiveness of the female body and spirit while in this state; the antipsoric medicine therefore acts more definitely and perceptibly during pregnancy. God be praised, the homeopathic physician who is acquainted with the natural means of a cure has little need for toxic drugs that weaken the body."

Young Animals

With young calves, *Aconite* often prevents the outbreak of acute viral infections. Secondary medications are *Calcarea phos, Sulphur,* E. *coli* nosode and bovine calf pneumonia nosode. The selection of the second medication depends upon the disease state that may be threatening the herd. *Calcarea phos* is chosen when there are teething problems or joint deformities like crooked legs. *Sulphur* is chosen when young animals appear stunted or emaciated, or when another carefully selected medicine fails to act. E. *coli* nosode is often useful when diarrhea or other enteric disease becomes clinical early in life. Bovine calf pneumonia nosode may be administered when clinical respiratory symptoms are manifest in one or more calves. The nosodes act more powerfully in prevention than in treating sick patients.

Constitutional Types: Which is Your Cow?

The Holstein dairy breed and the heavier beef breeds would benefit from the constitutional remedy *Calcarea carbonica*. These types of cows have large bodies, joints and hooves which make them move slowly. Because of their greedy appetite and a healthy thirst, they will tend to be overweight. There is a tendency to be prone to mucoid symptoms, arthritis and mastitis. Wet and cold weather and concrete will make these cows uncomfortable.

The Guernsey, Dutch Belted and Ayrshire cows with their active, athletic and lean bodies would respond to a constitutional dose of *Calcarea phosphorica*. With their aggressive appetite, *Calc phos* cows have a tendency to manifest illness via the gastrointestinal system with symptoms of colic, diarrhea, constipation and even bloat. They may show irregular cycles with failure to conceive. When anxious, they often produce frequent large quantities of manure. *Calc phos* calves drink milk but do not seem to thrive. *Silicea* should be considered for those animals that are prone to structural problems as with hooves or horns.

Holstein breeds may also benefit from *Sepia* and *Silicea*. Mentally the *Sepia* animal is very irritable but not mean, burned out, saggy, dragged down with poor body tone and worn out from one too many births. She may have silent heat or infertility with a tendency for uterine prolapse and mastitis. Giving *Sepia* on day 21 postpartum results in improved conception rates and shorter calving intervals. Research by Dr. A.V. Williamson and others from 1987 through 1990 utilized the use of 200C *Sepia* given postpartum for the prevention of anestrus. The results of these double-blind studies indicated that giving *Sepia* on day 21 postpartum increased conception rates and lowered culling rates.

The constitutional *Silicea* cow tends to be thin with dull hair and brittle or deformed hooves. Lacerations and other injuries often become infected, taking a long time to heal. The formation of scar tissue in the reproductive tract or udder is common. Using the homeopathic medicine when indicated during the dry period will help prevent fresh cow mastitis. Horses, cattle and goats who need the benefits of homeopathic *Silicea* tend to be thin with crumbly hooves and have poor feed utilization.

Graphites animals are heavy and slow moving with constipation as a frequent complaint. The skin, hooves and mammary gland can become thickened, cracked or scarred. Moist eruptions may occur between the toes or where the skin folds. The mammary gland becomes heavy with weakened and stretched suspensory ligaments. Nipples may crack and present a smelly discharge. The cow that responds to *Calc carb* will respond to *Graphites* when *Graphites*-type eruptions occur.

Jerseys would fit the *Pulsatilla* picture. With their sweet pretty faces and long eyelashes they start to walk toward you but retreat just before being touched. Their affectionate side can also be balanced by a grouchy side especially during times of maternal protectiveness. Friendly but timid with silent grief, they have a strong desire to stay with the herd, with their peers. When disease is manifest it will likely affect the mucus membranes producing thick, bland, yellowish-green discharges. Their symptoms will improve when out of doors and moving about and will be worse from being in the warm stable and eating rich feed. *Sepia* could also be incorporated for the Jersey since she has a tendency to "sag" faster than other breeds.

Within each herd there are cows whose overall symptom/personality picture may fit other polycrests such as, *Arsenicum, Lyco-podium, Natrum muriaticum,* or *Phosphorus.* The "boss" cow in any herd would fit the picture of *Phosphorus.* Bright, alert and outgoing, they drink volumes, milk volumes and urinate volumes.

Typical fears of the *Phosphorus* constitutional type include loud noises, thunderstorms and being alone. Like *Natrum mur,* this element is found in every organ and tissue of the body. There is a tendency for clotting defects and bleeding disorders. *Phosphorus* cattle stay far from strangers and the veterinarian. Horses that have a *Phosphorus* constitution love being watched and massaged. Because this polycrest has so many indications, it is advisable to keep *Phosphorus* 30C and 200C in the medicine chest.

Clean and fastidious—the cleanest cow in the herd fits the *Arsenicum album* type. They are nervous, easily excitable, fearful of the unknown and of approach from the rear. Easily chilled they are prone to respiratory allergies and/or diarrhea.

Lycopodium clavatumis is one of Hahnemann's great constitutional medicines. The patient is mentally strong but weak in the upper body. Prone to affections of liver, lungs, urinary and gastrointestinal tract, they will often bloat and have distention after meals. All symptoms are better with warm food and drinks. but are worse from 4 p.m. to 8 p.m. In its ordinary crude form, *Lycopodium* has almost no medicinal value on the human body. But when triturated and succussed, it provides the homeopathic practitioner with one of the great polycrests—a drug of many uses for animals and man.

Originally, Hahnemann used ordinary table salt to prepare the first doses of potentized *Natrum muriaticum*. No one until Hahnemann thought to research and look for the medicinal virtues lying hidden within ordinary table salt. Patients with a tendency for heatstroke, headaches, sinusitis or edema tend to respond to *Natrum mur.* Just as salt holds water, so these patients hold onto their dis-ease both physically and otherwise. A cow with cystic ovaries with frequent estrus, air in the vagina and an inflamed cervix would benefit from *Apis mellifica* 30C twice daily for three days followed by *Natrum muriaticum* 30C or 200C once daily for seven days.

Constitutional remedies for lactating animals are best given when the animal is confirmed pregnant. With routine herd fertility checks and management evaluations, the veterinarian and farmer will be able to evaluate symptoms and therefore select the appropriate remedy for that particular animal. The potency of a constitutional remedy can be any potency depending upon the age, present health status and previous health history. In a large commercial herd, giving each breed the matching constitutional remedy would be a simple way of incorporating the homeopathic medicine into the herd.

6.

Mastitis & Fertility-Related Problems

Mastitis

Mastitis is inflammation and infection of the mammary gland of all milk-producing mammals. It is caused by entry of bacteria into the teat canal and invasion of the tissue of the milk-producing glands. Bacteria are ever present in the environment of a dairy barn and are only waiting for suitable conditions to present themselves to invade tissue and set up infection. The advent of antibiotics was seen as the answer to this condition, as well as many others, but having used a multiplicity of antibiotics and combinations thereof, mastitis is as big, if not bigger, a problem today as it ever was. Antibiotics have been substituted for good hygiene and soon we had the emergence of antibiotic-resistant bacteria and an upsurge of virulent forms of mastitis, which did not respond to conventional treatment.

Acute Care

Acute episodes call for natural prescriptions. A hard, painful udder with a moderate fever indicates the remedy *Phytolacca*. Dosing three times daily for three to five days often is curative. Keynote symptoms are pain radiating from breast to other parts of the body (the cow tries to kick or get away), swelling of glands, and frontal headache with desire to clench teeth together.

If mastitis occurs early in the lactation period and is accompanied by an increased temperature, hot milk, dilated pupils, dry mouth, sensitivity to light and jarring, give *Belladonna*. Dose 200C every hour for two to five doses. Follow this with *Calc carb* if the udder is hard and firm with less sensitivity. Choose *Sulphur* if the temperature is lower, but the milk still has flakes. The second medicine may be given morning and evening until the symptoms are gone (usually in one to five days).

Persistence Pays Off: A Case Study

A Jersey cow with acute mastitis was treated twice daily with *Belladonna* 200C followed by stripping out of the affected quarters. Little improvement occurred. The farmer decided to "do it right and give the medicine as Sheaffer had said to do." He returned to the barn and spent the entire evening giving the *Belladonna* every hour along with hourly stripping of all affected quarters. By midnight the cow seemed quite a bit better. The temperature had dropped from 106 degrees to 104 degrees. With this improvement noted, the farmer went to bed. At 5:00 a.m. the cow was eating and 90 percent improved. In the days ahead she went on to make a complete recovery.

An extremely painful quarter in an irritable animal with purulent secretion indicates a need for *Hepar sulph,* which is best dosed three to four times daily. The pain worsens from touch, pressure, chill and cold drafts of air.

The mastitis case requiring *Aconite* would be characterized by a patient who is restless, fearful, worse at night, thirsty, and may have an accompanying cough. The temperature may range from 103 degrees to 106 degrees. You may use 30C or 200C every hour for up to four doses. Follow with *Sulphur* when the milk contains flakes and the udder is still red. When the temperature is in the 102-degree range, and flakes and glandular hardness persists, consider *Calcarea carbonicum*. Follow with *Nux vomica* when there is constipation and the appetite is depressed, or choose *Lycopodium* if ketosis is present, manure is irregular, and there is sensitivity in the liver region.

Apis mellifica from the honeybee is useful for cases of mastitis either before or after parturition. The breast is hard and swollen with edema. The cow prefers cool or cold applications rather than warm and likes shade rather than sunshine. Like *Pulsatilla,* she has decreased thirst and wants to avoid warm, stuffy buildings. One farmer reported to us, "When a fresh cow comes from the pasture with a hard, hot quarter and flakes in the colostrum, one dose of *Apis mel* is usually all that is needed."

If the mastitis occurs after freshening over the course of the day, and the patient is not thirsty, consider *Pulsatilla.* Keynotes for *Pulsatilla* are thick, creamy discharges with a timid, emotional personality. One farmer, after attending a homeopathic farm lecture, decided to "see if this stuff really works." After freshening, every cow received a dose of *Pulsatilla.* Every cow with retained placenta was dosed twice daily until it passed. Every cow with mastitis received *Pulsatilla* three times daily for two days. Every anestrus cow was given one dose daily until she began to cycle. Every sick calf with a thick, creamy discharge from the eyes or nose was also dosed with *Pulsatilla.* The results convinced him and he was converted to the use of homeopathic medicines.

Coliform Mastitis

Life threatening situations like acute mastitis and especially coliform mastitis cases demand more frequent dosing of natural medicines and more frequent initiation of support therapy such as stripping of the affected quarter. Symptoms of coliform mastitis may include: cold skin, droopiness with head held at half-mast, fatigued muscles and hard quarter. The udder may be hot or cold and it is usually painful. The milk may be watery or watery with red-brown blood or it may look and smell like calf scours. Usually not much can be milked out. From the farm kit, the remedy *Lachesis* and *Pyrogen* should be considered. If the discharge is watery and offensive, choose *Lachesis* 30C four times a day for two to three days. If the cow has a rapid pulse with a low body temperature or a feeble pulse with a hot body temperature, choose *Pyrogen* 200C three or four times a day for three days. *E. coli* nosode once daily along with the other homeopathic medicines may be instituted for the following five days, then once weekly for three to five weeks. As support therapy include B-complex orally or by injection once or twice daily, probiotic supplement twice daily, and vitamin C, 5-25 grams, twice daily. Hourly stripping and udder massage using Mastocream or *Phytolacca* cream should also be instituted.

Watery Mastitis

In cases of watery mastitis, administer *Pyrogen* 200C every eight hours for three days. Massage udder with *Phytolacca* ointment and the individual teat with *Arnica* ointment.

Gangrenous Mastitis

With gangrenous mastitis in lactating animals, alternation of the remedies *Lachesis* and *Pyrogen* have proven to save many animals. Because of the intensity and seriousness of the situation, these remedies are repeated every two or three hours. The best potency for *Pyrogen* in sepsis and toxemia is 200C or higher. The potency choice for *Lachesis* would be 30C or higher. As in other mastitis cases, institute the complementary measures of stripping the affected quarter hourly, massaging with Mastocream or *Phytolacca* ointment and administer systemic vitamin C at high levels and probiotics twice daily.

Septic Mastitis

An animal may experience severe septic mastitis, which is characterized by a rapid pulse with low temperature or a slow, feeble pulse with a high temperature. Discharges, including mastitic milk, are dark and offensive. The treatment history of the animal may have included antibiotic use such as dry cow mastitis tubes. Dose with *Pyrogen* 200C or higher three times daily for two or three days.

Supportive therapies could include:

1. Frequent striping of affected quarters to remove abnormal milk.
2. Fasting, reduce grain and many other high-protein or high-fat feeds until recovery.
3. Dose with the simillimum—the best matched homeopathic medicine.
4. Give antioxidants and probiotics at the first sign of inflammation. Examples of these are:
   a. Vitamin C, 5-25 grams twice a day for a full-size cow.
   b. Viable acidophilus cultures
   c. B-complex vitamins
   d. Electrolytes if the cow will take them

Time Is Required

Additional amounts of time are required when we gain anything new—whether it be relationship, tool, horse or medicine. Time is needed to observe. Time is needed to study. And time is needed to practice the knowledge gained. Those of use who daily care for livestock have a choice. The old saying goes, "A stitch in time saves nine." Using the homeopathic medicines now will save having difficult problems down the road. We make a choice each time an animal is ill. Between episodes of illness, we should be thinking prevention. A well-chosen homeopathic medication given at the right stage of lactation and during the dry period, results in fewer new cases and lower SCCs during the next lactation. Take time to give information to other farmers. Learning from each other is one of the great benefits of homoeopathic medicines. I knew a farmer who had a cow who had freshened two days earlier and now had bloody milk in the right front quarter with only minor udder swelling. She was off her feed with a temperature of 103 degrees. His good friend, who was also a farmer, suggested *Ipecac*. The medicine was given to the cow three times daily for two days and the milk cleared. The medicine resonated, vibrated, harmonized with the cow's unbalanced condition and the immune system responded by eliminating the illness. This is how we can help one another by sharing what we know.

Initiate Changes on the Farm

Initiate changes to eliminate cow-to-cow spread of infectious bacteria. According to Christopher Day, MRCVS, "*Streptococcus uberis* and *Escherichia coli* are always present in the barn and in the herd. They can *never* be eliminated from the barn. *Streptococcus dysgalactiae* and *Staphylococcus aureus* are present in carrier quarters (cows retain these bacteria in their udder) and in skin and also can *never* be eliminated totally from a herd. *Streptococcus agalactiae* is spread cow-to-cow during acute outbreaks and if properly treated, can be eliminated from a herd."

Staphylococcal infection is the most common pathogen involved in producing a high cell count response in a herd, and it is the organism least susceptible to being killed with antibiotics. For this reason, dry-cow antibiotics were developed, so that the antimicrobial drug could stay in the udder for a longer time and increase the kill rate. Some preparations are designed to persist for eight weeks. The dangers of this are obvious:

1. Foreign material is retained in the body for a long period, thereby increasing any potential toxic hazard to the cow.
2. The long-term presence of an antibiotic could theoretically increase the risk of inducing antibiotic-resistant bacterial populations.
3. Contaminated milk could enter the human food chain.

Disease and disease-causing bacteria are being suppressed or forced into deeper organs in the body. Most farmers with exemplary programs and high milk quality tell us, "When I stopped using antibiotic tubes for dry treatment, my overall herd health took a giant step forward." Following with well chosen homeopathic medicines will keep the herd moving in a positive, healthful direction.

Suggestions, with acknowledgment to the work of Dr. Christopher Day and his book, *The Homeopathic Treatment of Beef and Dairy Cattle,* to eliminate cow-to-cow spread of infectious bacteria would include:

1. Milk mastitis cases as a separate group at the end of each milking period.

2. Use a separate claw piece and bucket unit to plug into the vacuum line for milking mastitis cases.
3. Invest in one-way liners or claw pieces. This will reduce the reflux effect of milk which enters the udder from the claw piece. In normal systems, milk can surge back into the udder during milking, or residual milk can surge into the udder of the next cow onto whom the cluster is applied.
4. Use rubber gloves when milking to save hands from chapping or cracking. Cracked skin can be a source of infection, particularly from *Staphylococci*.
5. Use testing to identify carriers. Carrier cows can then be singled out for intensive treatment with homeopathy; they can even be isolated and milked separately.

Other suggestions include:
1. Review the entire nutritional program on your farm.
2. Make sure that protein is not too high in the total ration.
3. Eliminate molds from stored feeds, especially silage. When it is impossible to eliminate all molds, feed clay and charcoal to absorb them and give homeopathic medicines to cleanse the liver.
4. Daily exercise, sunshine and greens are important all year.
5. Test and quarantine all new animals.

Treatments

Treatments with nosodes and specific homeopathic medications will boost a cow's resistance; however, we should also be concerned with methods of actually reducing the bacterial invasion levels. Our current reliance on antibiotics has allowed the development of certain management practices, which would of themselves not be enough to induce resistance in the lactating animal. Relying on drugs is an approach to prevention somewhat like locking the barn door after the horse has been stolen. We should not wait until tragedy has occurred to begin plans to avert it.

Natural medical methods of treatment and prevention should be utilized throughout the lactation cycle. Treatment of the mastitis case starts before conception. Constitutional prescribing for both parents will produce healthy offspring. Once monthly dosing of a well chosen medicine during pregnancy is a good practice. Use these homeopathic medicines during the dry period according to the simillimum: *Calcarea carb, Magnesia carb, Silicea,* or *Calcarea phos*. Mastoblast at drying off has been used by quite a few dairy farmers with success.

Mastoblast and Mastoblast H.P. are homeopathic combinations manufactured in the United States and Ireland for the control of mastitis in organic lactating animals. The original formulation, Mastoblast required a 2cc liquid dose given orally. The H.P. formulation recommends a .5cc liquid dose, which may be administered orally, nasally, or in bulk drinking water.

The H.P. formulation is in wide use on organic dairy farms. It acts to stimulate a natural healing response in the body, in contrast to classes of conventional drugs in veterinary medicine, which can be immunosuppressive when used improperly. Some of the classes of drugs acting suppressively are: steroids, antibiotics, anticonvulsants, anesthetics, antihistamines, tranquilizers and vaccines. Ideally, agriculture should be carried on without drugs. However, there are times when treatments are required and homeopathic preparations like Mastoblast H.P. give the farmer and veterinarian new freedom.

There are three times when homeopathic combinations like Mastoblast H.P. are indicated: when lactating animals are subject to stress, when herd somatic cell counts exceed 75,000, and when there is clinical mastitis present in individual females or in the herd. The medicine is indicated for pregnant heifers and cows, drying off cows, and all lactating animals. Other species using Mastoblast H.P. are equine, caprine, ovine, camelid, porcine and canine.

General dosing should be a.m. and p.m. for the last five days of milking. Individual medicines are often dosed every 14 days during the dry period for three months. A well chosen milk nosode will complement the simillimum.

Inherited Tendencies

Inherited tendencies toward mastitis or other chronic diseases should be avoided when choosing breeding stock. If an animal has a history of mastitis or other problem, such as premature labor, retained placenta, metritis or infertility, it would be best not to continue to breed her. The chronic disease state causing the problem will continue to manifest itself unless a constitutional homeopathic prescription is given. Treatments with energy medicine is the best course of action. At times, even homeopathy cannot rid the animal of the deep-seated problem. Nutritional supplementation may have an effect in treating the problem in the present generation. Balanced nutrition, plus energy therapy and careful breeding are necessary to bring about the deep-seated healing that is needed for the present and future generations.

Sanitation

Sanitation means different things to different people. To all of us it should mean keeping the animal, the equipment, the milking station and the environment as clean as possible. A relatively new emphasis in the United States involves wearing milker's gloves. Human skin makes a wonderful vector for carrying certain mastitis-causing organisms from one lactating animal to another. Durable milker's gloves can be cleaned between every cow with boiling water or hot sanitizing solution. Whitewashing the barn annually was standard procedure—it was mandatory in the past. A farmer needed a certificate posted in the barn in order to ship milk. Why? The resistant organisms increase on site and can stay in wood and wet areas for years. Once yearly thorough cleaning and whitewashing of barns and milk stations should still be required.

Cows need daily exercise. Walking on pasture or sod and not in mud wallow is preferred. Forty years ago, our dairy cattle played and exercised outside every day of the year, even in snow. Only when there was a tremendous storm did the cows stay in the barn.

Somatic Cell Count: A Case Study

In a 40-cow certified organic purebred herd, eleven cows had somatic counts exceeding 5,000,000. Two others were experiencing active clinical mastitis. Milk cultures of the individual quarters and the bulk tank indicated which bacterial organisms were present. While waiting for the bacteria to grow so that a nosode could be prepared, the entire herd was treated with Mastoblast twice daily for 10 days. This treatment alone lowered the herd SCC from 1,100,000 to under 300,000, and bacterial counts dropped into the excellent range for the entire next month. Convalescent therapy consisted of prescriptions of *Phytolacca* and the herd milk nosode, which has maintained the health of the cows and the quality of the milk. The farmer also instituted many of the procedures for reducing cow-to-cow contamination.

To create a herd milk nosode milk samples are collected from a representative number of cows in the herd. A bulk tank composite sample is taken as well. Each milk sample is cultured on media to encourage growth of bacteria. When the growth reaches its strongest point, the bacteria are swabbed and the swabs mixed with a drop of milk. The mixture (solute) is stored in a small vial of 50 percent distilled water and 50 percent grain alcohol (100 proof vodka is an adequate solvent). The infused 50/50 mixture serves as a *mother tincture* for preparation of a specific milk nosode for that particular farm.

Reproduction and Related Problems

Utilizing homeopathic medicine in a dairy or beef herd can begin at any time. Cows that are not showing heat or estrus can benefit from a course of *Pulsatilla* 30C once daily for seven to ten days. Most have a 21-day cycle. Those with silent estrus can be given *Sepia* 30C daily for one to five days after ovulation and diapedesis (the outward passage of blood through intact vessel walls). Then on day 14 of the cycle begin with *Pulsatilla* daily until signs of estrus occur. By that time, the animal should be in her natural cycle and showing a strong estrus.

Caulophyllum, made from blue cohosh, has been helpful in all potencies to prepare the uterus for labor and delivery. Higher potencies stimulate the uterus. It is useful for stillbirths. The farmer has the option to give the expectant mother 12C or 30C once daily for the last two weeks of gestation. Cattle that tend to go over their due date will benefit from *Caulophyllum* daily. Following delivery, if the animal has a retained placenta, *Caulophyllum* should also be considered. Dose three times daily after delivery to expel the placenta. *Sepia* and *Lycopodium* are compatible with *Caulophyllum.* When there is a retained placenta, three main remedies should be considered as follows:

1. *Caulophyllum* when the cervix is closing prematurely;
2. *Sabina* when discharge is bloody with straining and circulatory turmoil; and
3. *Pulsatilla* when discharges are thick and creamy.

Sepia improves the tone of the uterus and stimulates the liver and thyroid gland. It is used to prevent abortion. General dosing after ovulation is once daily for five doses, after a miscarriage the dose is once daily for seven days. To prevent miscarriage dose once weekly for up to two months.

For small, inactive ovaries *Cimicfuga racemosa* is useful in treating pelvic inflammation; *Iodium* is the choice for animals that appear to be shriveled or wasting; and *China officinalis* should be administered to cows that are in a thin or dehydrated state.

Prevention of cervical and vaginal scar tissue can be accomplished through the judicious use of *Calendula* topically applied. Other remedies are *Silicea* for softening scar tissue, and *Graphites* for fibrous scarring.

Calving Paralysis

According to the *The Merck Veterinary Manual* (1991), "Calving paralysis or obturator paralysis is paresis (or paralysis) of the adductor and caudal thigh muscles of the hind limbs. The condition is a result of intrapelvic damage, primarily to the ventral branch of the L6 spinal nerve, a major contributor to the obturator and sciatic nerves; direct obturator nerve damage may also occur. These lesions are most frequently associated with dystokia, with signs of paralysis, paresis, or ataxia of one or both hind limbs. The condition is most common in cows, but other species also may be affected. Nerve injury occurs when the fetus lies in the pelvic canal for an extended period, or when a large fetus is forced through the pelvic canal." Homeopathic treatment choices would include: *Arnica, Hypericum, Conium mac.* Supportive therapy such as fresh dry deep bedding, turning the cow from side to side every few hours, probiotics and/or antacids to maintain digestive function, good quality hay and water every two hours, and if no evidence of hypercalcemia exists, the calf may be allowed to nurse.

Routine Care Procedures

Many of my clients routinely do the following treatments for their animals with good results. These simple procedures clinically provide the animals with a tonic effect and they appear to recover well from the calving experience with less of a tendency for a retained placenta and mastitis. The treated calves clinically put on weight and incidents of bone, joint and tendon deformities are reduced.

First-calf heifers are given *Caulophyllum* 30C once daily the last week of gestation. This prepares the bovine body, especially the reproductive organs, to give birth.

After calving give *Pulsatilla* 30C (20 pellets) in a bucket of warm water. Allow the cow to drink freely. *Pulsatilla* clears congestion of the womb.

Calves are given *Calc phos* 12X once daily for one week to prevent digestive and respiratory difficulties.

The Veterinary Examination

In all of the health states of the animal, the examining veterinarian needs to proceed unhurriedly and with an open mind. The animal caretaker needs as much information as possible about the internal state of the cow. They are relying on the veterinarian to give answers to questions. Are there cysts? Which ovaries? Are the ovaries of normal size? Is any part of the reproductive tract painful to pressure or touch? Is it hanging over the pelvic brim? Is there any discharge? Has the examination released any discharge? Is there scarring present in any tissue or organ? What is the general tone of the uterus?

Hurried, inadequate examinations lead to the prescribing of incorrect medications. Incorrect prescriptions miss the target so that little improvement is observed. In cattle, a herd health exam is both internal and external. The procedure should be gentle, careful and deliberate. As always, findings should be recorded and comparisons made with notes from previous exams to lead the farmer and/or veterinarian to the correct prescription.

7.

Specific Conditions & Remedies

Abomasal Displacement

Is displacement surgery always necessary? Talk to ten farmers who use homeopathy and you may receive ten different answers. Farmer A's best Holstein developed a displacement two days after calving. He wanted to avoid surgery if possible. Treatment consisted of alternating *Nux vomica* 30C and *Lycopodium* 200C every four hours. This treatment was continued for two days. The patient was bright and strong, but still off feed and the abomasum was still out of place. One dose of *Nux vomica* 200C was dosed on day three as a pre-operative medication. Surgery took place in textbook fashion with no complications. Post-operative therapy consisted of:

1. *Arnica montana* every eight hours for three days.
2. Feeding good quality mixed hay and avoiding fermented feed for at least 48 hours.
3. Offering just a few pounds of oats or corn chop three times daily.

4. Gradually increasing all feeds over two weeks until up to preoperative levels.

5. Suture removal 10-14 days postoperatively.

By day four following surgery, this cow was again up to her daily milk production of 120 pounds. Even though the twisted stomach was not corrected with homeopathic medicines, the cow experienced a positive response with the medications, which helped her through the surgical stress and convalescence.

At least one half of cows in organic herds are able to avoid surgery when treated with homeopathic medications. Ketosis is a frequent symptom in a cow with a stomach out of place that is not functioning properly. *Lycopodium*, *Nux vomica*, and/or *Phosphorus* will address ketosis energetically. Frequent dosing helps to restore liver function and normal peristalsis. Feeding feeds high in energy and high fiber feeds as well as dosing with B-complex vitamins and probiotics daily will complement the homeopathic medicines.

Arthritis

See Osteoarthritis

Bites or Stings

Ticks, spiders, mosquitoes, and other insects can cause illness from their venomous stings. Though not usually fatal, the bite of any creature may result in local inflammation or systemic illness (sepsis). With that in mind, we recommend that the following medicines be included in your homeopathic first aid kit:

1. *Ledum palustre*. Give three times daily. Occasionally, following vaccination, an animal will develop a local inflammatory reaction at the injection site. Administering *Ledum* both orally and topically reduces the pain and inflammation.

2. *Apis mellifica*. Taken orally, it decreases the reaction from the stings of bees, wasps, and hornets when ice brings relief.

3. *Urtica urens*. Orally or locally for stinging, burning pains which are often relieved by warm applications.

4. *Arsenicum album*. Taken orally; red and swollen eruptions of the skin which are better from warm or hot applications.

5. *Lachesis muta.* Wounds become purplish blue and bleed
   easily. There may be a toxic, systemic reaction to fleas.
There may be a toxic, systemic reaction to any animal venom.
Dosing orally in acute situations every hour for two to five doses
is recommended.

Bloat

Give *Carbo veg* 30C alternated with *Nux vomica* 30C dosing
hourly until symptoms improve. As with any other condition, if
the patient worsens, contact your local veterinarian immediately.
Prop up the animal with a bale of straw and place crosswise in her
mouth a heavy piece of rope. Chewing the rope increases salivation and enhances eructation (belching).

Adjunct therapy:

1. Offer more long stem hay before grazing.
2. Milk of Magnesia orally three times daily as directed on
   the label.
3. Vegetable oil drench may be given instead of Milk of
   Magnesia three times daily. Give mature cattle six to
   eight ounces per dose; calves one to four ounces depending upon size and age.

Bloat: A Case Study

A herd of 40 Holsteins were grazing on fresh alfalfa
on a June day. When bringing in the herd for milking, the
following symptoms were observed—bloat and distention,
shortness of breath, loss of appetite and staggering gait.
After a quick call to her holistic vet for advice, the farmer
began to treat every cow with *Carbo veg* 30C orally. Half an
hour later she gave every cow a dose of *Nux vomica* 200C.
All but three cows responded well, and those three were
given a second dose of *Carbo veg*. Cost of treatment—$26.00
for medicine, plus one hour of labor.

Bovine Foot Rot

Bovine Foot Rot is known by the following synonyms: podo-dermatitis, Paronychia, foul-in-the-foot, and *Fusobacterium necrophorum* infection. It is a painful inflammation found also in the other cloven-hooved species. *Fusobacterium* is implicated in cattle and related organisms have been cultured from sheep and goats. Generally, the soft tissue below the pastern or between the hooves is found to be swollen and red. There may be pockets of pus in the claw, in the cleft or around the coronary band.

A conventional therapy of trimming the fibromas and necrotic horn with bandaging is helpful, but only goes so far toward a cure. Homeopathic therapies will stimulate healing deep in the tissues while moving any infection to the surface of the skin. Three medicines to be used in early stages for bruising are *Arnica montana, Bellis perennis,* and *Hypericum perforatum.*

If the condition has progressed untreated for several days, *Hepar sulph* would be indicated in alternation with one of the above medications. For example, when there is a puncture of sole, heel or interdigital space *Hypericum* in alternation with *Hepar sulph,* dosed about every four hours, brings relief and tends to move any infection from deep tissues to the surface.

Foot Rot in later stages always requires some surgery to remove diseased and necrotic tissue. The lower leg and foot is bandaged after surgery with *Hypericum/Calendula* ointment or other antiseptic dressing. The homeopathic pellets for oral dosing in later stages, depending on the total symptom picture (see *Materia Medica,* Chapter 8), include:

1. *Myristica*—dose TID for three to five days,
2. *Silicea*—dose once daily for seven to 14 days,
3. *Graphites*—dose BID for seven to 10 days, and
4. *Kreosotum*—dose BID for seven to 10 days.

Bovine Foot Rot: A Case Study

Three cows in a herd of Jerseys developed symptoms of foot rot with lameness, swelling and lesions. The local conventional veterinarian cautioned that the entire herd would probably all become infected. The in-state homeo-

---

pathic veterinarian suggested dosing the entire herd with Bovine Foot Rot Nosode, a product produced by Ainsworth Homeopathic Pharmacy in London, but available by prescription only from homeopathic pharmacies in the states. Using 15 gallon tubs, the entire herd was dosed three times daily for three days. The active cases were treated with *Hepar sulph* in addition to the nosode. Three days after the last dose of nosode there were no active cases and all previous cases had recovered. One month later, a milking Shorthorn bull was brought on the farm for breeding purposes. After receiving permission from the owner of the bull, the animal was treated with the nosode. While on the farm for two months, the bull remained free of Foot Rot. When the Shorthorn bull returned home, a "clean-up" Jersey bull was brought on the farm. On-farm demands of time and energy prevented the Jersey from being treated with the Foot Rot nosode and the Jersey bull developed signs of Foot Rot infection.

Bronchitis

See Respiratory Affections

Bruises

Arnica montana is useful when bruising occurs. Dosing three or four times daily for three days is recommended. *Bellis perennis* made from the English daisy is useful when there is repeated bruising of the sole, udder, or the perineum. This remedy follows *Arnica* well and helps prevent fibroids and tumors at the location of the bruise. Dose twice daily for a week with 12C or 30C potency at the first indication of hardness in the tissues. Topical ointments, creams and gels that contain *Arnica montana* and *Bellis perennis* are now available from several homeopathic pharmacies. Apply topically three times daily.

Calf Diphtheria

See Respiratory Affections

Calving

One has the option to give the expectant mother *Caulophyllum* 12C or 30C once daily for the last two weeks of gestation. Following delivery, if the animal has a retained placenta, *Caulophyllum* should be considered again. If the retained placenta is the result of the calf's abnormal position, consider *Pulsatilla,* and if it is accompanied by bright red bleeding, give *Sabina* 30C or 200C three times daily is recommended. See also chapter on Mastitis and Fertility-Related Problems (Chapter 6).

Calving Paralysis

According to the *The Merck Veterinary Manual* (1991), "Calv-ing paralysis or obturator paralysis is paresis (or paralysis) of the adductor and caudal thigh muscles of the hind limbs. The condition is a result of intra-pelvic damage, primarily to the ventral branch of the L6 spinal nerve, a major contributor to the obturator and sciatic nerves; direct obturator nerve damage may also occur. These lesions are most frequently associated with dystokia, or dystocia, with signs of paralysis, paresis, or ataxia of one or both hind limbs. The condition is most common in cows, but other species also may be affected. Nerve injury occurs when the fetus lies in the pelvic canal for an extended period, or when a large fetus is forced through the pelvic canal." Homeopathic treatment choices would include: first day: *Arnica* 30C or 200C, four times daily; second day: alternate *Arnica* and *Hypericum* 200C, four times daily; third day: alternate *Conium mac* 200C and *Hypericum* 200C, four times daily.

Supportive therapy:
- Fresh dry deep bedding.
- Turn the cow from side to side every few hours.
- Probiotics and/or antiacids to maintain digestive function.
- Offer good quality hay and water every two hours.
- If no evidence of hypercalcemia exists, the calf may be allowed to nurse.

Chalazion

"Chalazion is the medical term for a granulomatous inflammation of the glands of the eyelid. It is similar to a stye, but it usually does not come to a head and is more difficult to eradicate. *Platanus* is my chief remedy for Chalazion in the eye. I instruct the patient to take a solution of *Platanus* and *Calendula* tincture 50/50, dilute one drop in a tablespoon of water, and put in the eyes three times daily. It doesn't work instantly, but if you use it for several months, the lesions start to recede. If there is any hint of recurring inflammation, start using the drops again. I've never seen anything work as well." (David Wember, M.D. in a speech before HMSSP, February 13-15, 1998, Cocoa Beach, Florida.)

Coccidia

Coccidiosis is a disease of unsanitary conditions; it becomes clinical when calves or lambs or kids are reared in the same pens and pastures year after year. It is common during the warmer months if livestock are confined. Proper pasture rotation will greatly decrease the number of infective organisms that are available to animals. These protozoan parasites of various species of the Eimeria family are passed via the manure from as few as one animal onto the herbage (grass, clover, alfalfa) and soil, where part of their life cycle is spent maturing into infective oocysts. When infective oocysts are ingested by other animals, they are subject to contracting the disease.

First and foremost, farmers need to get into the habit of rotating pastures and pens. This simple management tool breaks the life cycle of protozoa just as it does for intestinal helminthes. A harsh winter will go a long way in cleaning up pastures, but stalls and pens must be thoroughly cleaned and disinfected several times a year. Since cattle over two years of age rarely are susceptible to Eimeria, adult cattle can occupy the pens where calves had been previously. Calf pens should be moved seasonally or quarterly.

Homeopathic therapies for treatment and prevention:

Ipecacuanha—Animals will have diarrhea with flecks of whole blood in the manure. They lose their zest for eating and the sight or smell of normal feed causes nausea and loss of appetite. Dose with 30C or 200C TID for five days.

Mercurius corrosivus—Animals have been ill several days with dark mucus in the diarrhea. The manure has an offensive odor and there is much straining. Dose with 30C or 200C BID for seven days.

China officinalis—Useful in early febrile stages or the later diarrheic stages. Animals will show weakness from loss of body fluids and dehydration. Dose with 30C TID for three to five days.

Sycotic Co. 30C—Drs. Paterson and Bach prepared this bowel nosode from an intestinal bacillus. As a homeopathic medicine, *Sycotic Co.,* has specific affinity for any inflamed mucous membranes, especially the intes-tinal tract during subacute or chronic diarrhea. For treatment of clinical cases, dose along with one of the previous medications once daily for three days. For prevention along with sanitation and rotation of pens, dose once weekly for five weeks.

Conjunctivitis

See New Forest Eye (NFE)

Colic

See Indigestion

Constipation

High forage diets with much lignified plant fiber along with various toxins will produce symptoms. When there is much strain-ing with little thirst, *Nux vomica* is strongly indicated. When the manure is hard and dry and the animal is very thirsty, *Bryonia alba* is the choice. If bloating is a consistent symptom, alternating *Carbo veg* with *Nux vomica* hourly usually brings relief. Laxative treatments such as digestible oils or magnesium oxide will encour-age expulsion of the highly lignified ingesta. If toxins or molds have been ingested, feeding bentonite and charcoal will act to absorb poisons.

Corns

See Bovine Foot Rot

Cystitis

See Urinary Tract Affections

Dehorning

The protocol for dehorning using homeopathic medicines is as follows:

1. At the moment of restraint, dose with *Aconitum napellus* to allay fear and anxiety.
2. As soon as surgery is completed, give *Arnica* for bruising, bleeding and pain.
3. Follow with *Hypericum* when local anesthesia has been employed.
4. For any unusual hemorrhage, the medicine usually selected is *Phosphorus*.
5. When environmental temperatures exceed 90 degrees, delay surgery.
6. If the barn is hot and poorly ventilated, change location.
7. Surgical sites should be medicated with *Hypericum*, *Calendula* or *Hypercal* daily to speed granulation and deter insects.

Dentition

See Stomatitis

Diarrhea and Dysentery (Enteritis)

When confronted with diarrhea there are a number of choices regarding particular medicines. This is a case when careful observation of the symptoms is important. The first medicines that come to mind are *Arsenium album, Carbo vegetabilis* and *Podo-phyllum pelatum*, the choice would be made by observing the discharges and other symptoms the patient is showing.

The keynote symptoms for *Arsenicum album* are burning or acrid discharges with nausea and loss of appetite. The muzzle, nose and extremities are cold and the patient is thirsty for frequent sips of liquid. Water and electrolyte water should be available to these patients at all times.

Carbo vegetabilis has bloating along with indigestion. The patient may have a cold muzzle and cold breath, but they still crave fresh air. If confined to a stuffy barn these animals are prone to collapse. They are better with a fan and in fresh air. Like *Arsenicum album*, the stools may be acrid and have a cadaverous odor.

Podophyllum peltatum from the May apple produces a watery profuse diarrhea in those animals affected by its poison. Other keynote symptoms are swollen glands, desire for cold water and the intense desire to press the gums together. You will notice cattle frequently biting on hard objects like fence boards and water pipes.

Phosphorous is a frequently used polycrest when cattle have eaten spoiled feed, particularly feed that is contaminated with toxic molds. The animals are very thirsty and have a fever or cough. *Phosphorus* follows *Podophyllum* and *Arsenicum album* well in enteritis.

Additional remedies are *Chamomilla*, which is particularly good for young, teething livestock; *Calcarea phosphorica*, if the illness follows a sudden change in weather; *Ipecacuanha*, if there is nausea not relieved by vomiting; *Mercurius corrosivus* when there is straining not relieved by stool; *China officinalis* for undigested, frothy stools with much gas and rumbling and pain with pressure; and *Sulphur* when other indicated medicine fails to act.

See also Indigestion.

A *Podophyllum* State: A Case Study

A herd of 42 Holstein dairy cows presented signs of watery diarrhea, depressed appetite, and rumen acidosis. The herd owner believed that it may be an outbreak of Winter Dysentery although the month was September and the weather was unusually mild. Both older cattle and first calf heifers were affected. Winter Dysentery generally hits two-year-olds and three-year-olds the hardest. In this outbreak, cattle with the largest appetites were most severely affected. After a thorough history and case review, it was noted that the owner had depleted his silage and had bought enough from a neighbor to feed until his own harvest in late October. Inspection of the purchased silage indicated two potential reasons for digestive upsets: small particle size and the presence of mold. The prescription was *Podophyllum peltatum* three times daily for three days. Reduce silage feeding by half and make up the difference with a mixed grass hay. The result was that 90 percent of the herd responded and were well in three days. Several stubborn cases were given *Phosphorus* twice daily for three additional days. All had an uneventful recovery.

Eyes—Bruising or Blunt Trauma

When the cornea has a bruised appearance, consider *Symphytum officinalis*. There are times when a veterinarian examines an eye and is not able to determine if the inflammation is caused by infectious agents or blunt trauma. *Symphytum* three times daily is the medicine of choice. *Symphytum* has a similar uniting effect on the tissues of the cornea and conjunctiva as it has on bone with slow-healing fractures and on ligaments that have torn. *Ledum palustre* should be kept in reserve in cases of blunt eye trauma. In cases where *Symphytum* doesn't completely heal, a secondary medication is indicated. The use of *Ledum* will often complete the healing process.

Fractures

"What can homeopathy do with broken bones? It has been my fortune to see numerous cases of green stick fractures mostly, but not exclusively, in young subjects. In every instance the need for *Calcarea carbonica* was plainly indicated by the fat, lymphatic general appearance of the subject; his tumid abdomen; and sweaty head, neck and extremities. *Calcarea ostrearum* had never failed to help the injured member and to improve general health at the same time. I do not recollect to have given any besides the two hundredth potency," stated Spencer Carleton, M.D. in his text *Homeopathy in Medicine and Surgery* in 1913.

In bovine fracture cases, if the bone can be immobilized and the animal be given stall rest, the prescription of alternating doses of *Calcarea phosphorica* 30C and *Symphytum* 12C every four hours should be given for the first week. As the union progresses, the treatment would change to *Calcarea phosphorica* 30C once daily and *Symphytum* 12C once daily at different times of the day for an additional two weeks. Upon radiograph evaluation, if osteogenesis is progressing, the treatment regimen should change again to *Calcarea phosphorica* 30C once weekly and *Symphytum* 200C every other day. After four to six weeks, there should be strength and stability at the fracture site. If so, the patient may be turned out alone into a small paddock for grazing. All efforts should be made to protect the injured limb from future trauma. Once it is determined that the fracture is strong enough to withstand normal activity, the prescriptions are discontinued.

Grass Tetany

Hypomagnesemic tetany can occur in both cattle and sheep that are grazing rich recently fertilized pasture. Fertilizer and manure that is abnormally high in nitrogen and potash (potassium) will create the conditions to produce hypomagnesemia. The plant grows so rapidly that it is not able to take up adequate calcium and magnesium. In lactating cattle (when production is at the highest) dietary calcium and magnesium may not be able to meet demand for lactation and maintenance of good health.

The first signs are muscle twitching and stiffness of the limbs. If these symptoms are not recognized, and the livestock are not moved from the affected pasture the course of the disease may rapidly progress.

Providing calcium and magnesium rich hay before grazing may prevent illness and death. Having free choice minerals available to the herd at all times is another lifesaving approach. Make sure at least one bin contains high magnesium minerals, another has high calcium, and a third contains available phosphorous. Free choice Real Salt or sea salt should be available in a fourth bin.

There are times that the grass is so imbalanced and deficient in magnesium and calcium that the livestock must be removed. In that case, hay can be harvested and the bales sprinkled with salt and magnesium oxide or magnesium sulfate.

After the initial muscle spasms, symptoms of the acute disease include throwing the head back and falling down on their sides with paddling convulsions. At this time IV (intravenous) calcium and magnesium solutions may save a life. Intravenous solutions must be given slowly and carefully while monitoring the heart. The cattle should never be startled or prodded with a 'hot stick', but should be allowed to rise at their own speed.

The following homeopathic medications will complement the patient:

Magnesia phosphorica 30C. When the first signs are observed, keynote symptoms such as muscle twitching and cramps are observed. The homeopathic medicine *Mag phos* will signal the body to conserve magnesium. This medicine may be dosed in alternation with any of the following.

Belladonna IM. The peracute phase exhibits sudden falling with tonic, clonic convulsions, dilated pupils and elevated temperature. Recommended dosing is every half hour or between convulsive episodes for up to six doses. When working around these animals, slow movements are very important; if sick animals are suddenly startled, they may die.

Cuprum metallicum 200C. Muscle spasms begin in the limbs and move proximately to the chest until the entire body is involved. There is constriction of the chest muscles that spreads to the abdomen leading to signs of colic with dark greenish diarrhea. Usually these animals fall down due to rigid limbs, dyspnea and colic without convulsions. This medicine given every one or two hours will complement intravenous therapy. Like the *Belladonna* patients, they are worse from touch and may die of fright. I have seen this sudden death syndrome in both cattle and sheep.

Kali phosphoricum 30C. There appears to be weakness to paralysis of several groups of muscles without the violent cramps or convulsions. It is recommended to dose every four hours in alternation with *Magnesia phosphorica* for up to five days.

Gelsemium sempervirens 200C. A convalescent medication to aid in restoring balance and strength is highly recommended. The patient is droopy, drowsy and trembling, acting as though she has head pain. This medicine has restorative action on the motor nerves and muscles. Dosing TID for three days will help insure full recovery.

As a preventative in pregnant cow and heifers dose with *Kali phos, Mag phos,* and *Calcarea phos* once weekly for the last two months of gestation. High nitrogen, high potash fertilizers are to be *avoided* on pastures. In pastured cattle, hypomagnesemic tetany has been known to occur in wet weather without the addition of fertilizer. The condition has been observed in beef cattle on silage that previously received high nitrogen, high potash fertilizer.

Hardware Disease

Also known as Traumatic Reticuloperitonitis, this illness occurs when ruminants ingest pieces of metal and other sharp objects that are mixed into their feed. It is up to the herdsman to keep paddocks and pastures free of nails, wire and other junk. Cattle have been known to eat bolts, screws, nails, a bag of concrete and even a croquet ball.

One step in prevention would be to give each yearling heifer and young breeding bull a smooth, powerful magnet to swallow. The magnet will attract all objects made from iron or steel. A good magnet should last at least three years. Every cow that is suspected of suffering from hardware disease should receive another magnet, especially if there is no record of a magnet being given during the last three years. Secondarily, if it is winter and cattle are being housed in tie stalls, extra clean, dry bedding should be placed under the front half of the body. Whether the cow stands or is lying down, her withers should always be higher than the hip bones. Pressure of abdominal contents on the diaphragm will be lessened when they take this position.

The following homeopathic medicines will be complementary and may reduce pain and suffering:

Arnica montana. Initial signs of pain and depression when vital organs are penetrated by rusty metal. Wounds to the rumen, reticulum, omasum, and abomasum are not as life-threatening as penetrations of the diaphragm, pleura, lungs, pericardium or heart. Injuries to these later organs could be fatal. Dose TID for three to five days.

Bryonia alba. Traumatic pleuritis often causes severe chest pain. Pressure relieves the chest pain so that she will be frequently found lying down on one side. If not lying down, the cow will stand very still, being reluctant to move. Usually there is fever accompanying extreme thirst. Dose TID for five days.

Hepar sulph. When contaminated metal migrates from inside the forestomach and penetrates other organs, purulent pleuritis or peritonitis may result. Symptoms of an elevated temperature and elevated white blood count along with chest pain are confirmatory. Treatment of 30X or 30C *Hepar sulph* TID may help any infection to drain to the outside. Use of higher potencies 200C and IM may remove infection via the circulation.

Silicea. The medicine of choice to be administered when foreign metal objects such as nails, screws and wires are not removed surgically. This medication may help to expel the offending object(s). *Caution!* Never prescribe *Silicea* when the patient has any prosthetic device in the body such as heart valves, artificial joints or hernia repair mesh.

Heat exhaustion

Administer *Belladonna* 200C every one-half hour for three doses followed by *Natrum mur* every one-half hour for three doses. If the animal acts like it has a headache, use *Glonoine* 30C or *Gelsemium* 200C every one-half hour for three doses. Adjunct therapy would include tepid water baths and moving the animal into shade.

Hoof Fistulas

Fistulas in hooves are a manifestation of chronic disease and the result of treading upon gravel and shale walkways. Weakness of the horn in the wall or sole allows foreign material to become embedded and further trauma or stress can lead to infections. Homeopathic medicines to consider include the following:

Hepar sulph 30X. Extreme lameness with pain; there is pressure and puffy swelling at the coronary band. The hoof may be hot and the animal prefers warm or hot soaks rather than cold. Dose three times daily until adequate drainage occurs. The opening may be flushed with *Calendula* mother tincture in hydrogen peroxide and saline several times daily. Combine 5cc of *Calendula* tincture and 5cc of peroxide with 50cc of isotonic saline. Ninety percent of the cases do not need bandaging. However, if lameness is severe or the sole has been eroded away, the hoof may be bandaged with *Hypercal* ointment or *Ichthammol* ointment. The bandage should be changed daily.

Silicea 30C. Useful after drainage nears completion. Patient is less lame and less painful to pressure. *Silicea* will help the body to expel tiny bits of foreign material and promote stronger hoof growth. Dose once daily for one to two weeks.

Myristica sebifera is useful in cases that do not respond to *Hepar sulph.*It is often called the "homeopathic knife." It has specific action on the tissues around the nail and leads to drainage of purulent material from hooves and joints. Choice of potency would be 12C or 30C given twice daily for five to seven days.

Arnica montana 200C. Use early when signs of lameness appears. With deep bruises of the hoof, apply *Hypercal* or *Arnica* ointment under a heavy bandage that should be changed every two or three days. A veterinarian or competent farrier will trim the damaged portion of the sole when the bandage is changed. Waterproof bandages and/or rubber boots may be used to protect the hoof until new horn growth fills in the defect. Dosing of *Arnica* 200C or *Ledum* 200C after trimming is recommended, dosing TID for three days.

Indigestion

Signs of indigestion are an atonic rumen with trapped gas and little motility. There is always depression with loss of appetite and there may be bloat. Rumen contents, which should have a basic pH become acidotic and turn into a doughy mass. Spoiled food or overeating of rich food is the usual cause. Cattle fed highly processed feeds or bakery waste are at high risk. Surgical intervention in the form of a rumenotomy to remove the acidotic and toxic contents, if done early enough, will often save lives. Antacids, laxatives, and probiotics are indicated orally three times daily. Other recommended medicines include:

Nux vomica. Constipation and straining with poor rumen motility are keynote symptoms. Manure is scant, hard and dry. Dosing hourly before surgery (if performed) should shorten recovery time.

Carbo vegetabilis. Bloat and toxemia are frequent symptoms with fainting or collapse as a sequella. Animals will have cold skin and cold breath. Manure, if passed, will be acrid and excoriating of rectal tissues. It is recommended to dose hourly for up to seven doses in alternation with *Nux vomica* until surgery is performed.

Colocynthis. Keynote symptoms are cramping pains with diarrhea and offensive flatulence. Calves stand with both hind legs pulled under their bodies. Adult cattle will lie down often on the same side with one leg drawn underneath the body. There is accompanying restlessness and agonizing abdominal pain. Each time the animal eats or drinks, she passes jelly-like dysenteric stools. The recommended dosing is every one to three hours as needed to control pain.

Colchicum. Ruminants calling for this medicine will have distention on both sides of the abdomen. Keynote symptoms are gas and rumbling with a tendency to collapse (like *Carbo veg*). Belching is difficult and manure tends to scanty, but not hard and doughy as with *Nux vomica*. The smell of food causes nausea or fainting and there is usually an accompanying stiffness of joints. Hourly dosing is recommended until distention is relieved.

Iritis

Iritis is also know as Uveitis. *Hepar sulph* is indicated when Iritis develops and there is pus in the anterior chamber just beneath the cornea. This would probably occur after the animals have been affected for several days, but have gone unnoticed and without proper medication. Dose three times daily for four or five days. Spray the face and eye with *Hypericum* saline three to four times daily as well.

To prepare eye wash, add five drops of *Hypericum* mother tincture to one ounce of isotonic saline.

Phosphorus in elemental form as a poison leads to degeneration of the retina, to cataracts, to glaucoma and to atrophy of the optic nerve. Homeopathic *Phosphorus* is indicated any time that a patient suffers from ocular inflammation accompanied by photophobia and increased thirst. In livestock, the sclera will be injected red or orange and injuries will bleed freely. The patient tends to be high strung and fearful. Recommended dosing is TID for five days.

Ketosis

If ketosis is a problem, a producer will need to search for a cause. Is there adequate quality fiber in the ration? Are the dry cows getting over conditioned? Is the cow eating moldy silage? Eliminate the mold and help the cow detoxify by using *Phosphorus* alternated with *Chelidonium*. One could give eight to ten pellets of *Phosphorus* in the morning and eight to ten pellets of *Chelidonium* in the evening until the symptoms are gone. If the cow has ketosis and is constipated, then *Nux vomica*, eight to ten pellets morning and evening, should be helpful. If she tends to be gassy and mildly bloated off and on, then *Lycopodium* will be the choice, dosing two or three times daily with the same amount of pellets. Feed good quality mixed hay and cut back on silage. Make sure she is getting plenty of daily exercise even if someone has to jog with her around the pasture. If there is ketosis and diarrhea, then consider *Phosphorus* eight to ten pellets twice daily. All the above medicines may be given in the 30X, 30C or 200C potency. In all ketosis cases, additional use of B-complex and probiotics will assist the animal in returning to a balanced metabolic state.

Laminitis

Laminitis is inflammation and edema of the sensitive lamina of the hoof. Also called Founder, it can occur in all hoofed animals. In the United States, it is seen most often in the equine and bovine species. The main cause in cattle is grain overload, but laminitis can also follow acute febrile diseases, acute indigestion, and certain toxic conditions. Wheat breaks down in cattle faster than corn and therefore, in cattle, wheat overload can cause lactic acidosis and laminitis faster than corn overload.

Signs of acute Founder are increased pulses in the extremities, especially the feet, pain and restless treading and a flattening of the sole. Secondarily, within days the veterinarian may observe early sole ulcers or white line disease. All cattle with laminitis, regardless of age, should be placed on a diet of grass hay and adequate clean water until recovery is complete. Treatment is as follows:

Aconitum napellus 12C, 30C or 200C. This is indicated at the first sign of inflammation. Dosing every three or four hours is recommended during the first day.

Nux vomica 30C. This should be dosed hourly in cases of grain overload. Other laxatives, fluids and electrolyte solutions will encourage emptying of the rumen. Surgery to empty the rumen should be considered if it can be performed within the first twenty-four hours. Some clinicians are adept at using a large bore tube and stomach pump to flush the excess grain from the rumen.

Rhus toxicodendron 30C. This is indicated for stiffness resulting from muscle spasms and increased tension on the deep flexor tendon. By day three, dosing in alternation with *Aconite* will decrease inflammation, relieving stiffness and improving gait. An example of therapy for day three, four and five would be *Rhus tox* at 7 a.m. and 3 p.m. and *Aconite* at 11 a.m. and 9 p.m.

Ligament Injuries

In all species, *Symphytum* is helpful for ligament tears and for shin splints (periostitis). It follows *Ruta* well in ligament injuries. Ligaments can be considered in a category with cartilage and bone and the same homeopathic medicines used for these symptoms would apply. Treatments with *Ruta graveolens* and or *Symphytum officianlis* are recommended twice daily for two weeks. Topical applications of ointments or lotions prepared from one of these two medicines will act in a complementary manner.

Ligament tears in young, growing animals will heal more quickly if given *Calcarea phos* daily during the first week of injury.

Luteal Cysts

Luteal cysts are frequently found when no estrus is observed. Medications often prescribed are *Apis mel* (twice daily for three days), *Natrum mur* (once daily for seven to ten days), *Calcarea phos* (twice daily for five days or once daily for ten days), and *PTT* (Posterior Pituitary, Thymus and Thyroid, daily for seven days). *PTT* is a by prescription-only product that can be purchased through a veterinarian. In the recent past, it has been manufactured by Arrowroot Standard Direct in Paoli, Pennsylvania.

Mastitis

See Chapter 6, Mastitis and Fertility-Related Problems

Milk Fever

One farmer treated with *Calc phos* 30C every 15 minutes. By the third dose, the cow sat up, and by the fifth dose, she was standing and starting to eat. Other farmers have had success with alternating *Calc phos* 30C and *Magnesia phos* 30C. If the cows are obese, alternate *Calc carb* 30C and *Calc phos* 30C. Prevention of milk fever may be accomplished by giving the cows with this tendency *Calc carb* 30C, *Magnesia carb* 30C, *Calc phos* 30C or *Magnesia phos* 30C every 14 days during the dry period. The Holstein dairy breed and the heavier beef breeds would benefit from *Calc carb* and *Magnesia carb*. With their active, athletic and lean bodies, the Guernsey, Jersey, Dutch Belted and Ayrshire cows would respond to the *Calc phos* and *Magnesia phos*.

Navel Ill (Joint Ill)

Sanitation at calving is extremely important. The best place for birthing large animals is a clean, fresh pasture that is in a rotational system. Calves that are born in stalls and lots must have the navel treated with tamed organic iodine or *Calendula* mixed with olive oil. The navel stump may be medicated several times on the first day of life if necessary.

When disease becomes clinical, the following medicines are useful:

Aconitum napellus 200C or IM. This should be given at the first sign of febrile illness when body temperature rises (103-105°F). Dose hourly for up to five doses.

Hepar sulph 200C or higher. There is purulent discharge from the navel and the illness has been progressing without homeopathic treatment for three or more days. Signs are fever, depressed appetite, lethargy and joint pain. Dose three times daily for three days.

Pyrogenium 200C or IM. High body temperature is accompanied by a weak pulse, or a low body temperature accompanies a firm pulse. The discharge from the navel is watery and offensive. The calf's vital force is low as it is in a weakened condition with a poor prognosis. Dose three times daily for three days.

Streptococcus 30C. Hemolytic bacteria homeopathically prepared has been useful when administered along with one of the above medications. Recommended dosing is BID for five days.

Nephritis

See Urinary Tract

New Forest Eye (NFE)

This ailment is often helped in early stages by *Aconitum napellus* and *Hypericum* saline spray. Later stages respond to *Euphrasia*, diluted tincture topically and the 30C potency orally. Chronic cases call for oral *Silicea* in potency. Here again, *Hypericum* sprays or compresses are helpful at all stages. *Hypericum* spray may be prepared by adding five drops of mother tincture to one ounce of water or 0.9 percent (isotonic) saline.

Prevention of New Forest Eye is augmented with the use of the nosode bearing its name. Often, this is prescribed in the spring, before the animals are turned out onto permanent pasture. Brush and shrubs contribute to eye irritation. Flies and other insects also irritate ocular tissues. The presence of viruses and bacteria in the herd play a role. Along with the NFE nosode, sound management will control the other factors contributing to inflammatory or contagious eye conditions. Treating a group of animals can be made easier by dosing the common water supply. Obtaining the NFE nosode from your veterinarian, dissolve 100 pellets (or pillules) in a gallon of clean water. This may be used as a stock solution for medicating drinking water. Use a fresh stock solution every 24 hours.

Oral Pain and Toothache

Oral pain and bleeding come under the sphere of *Hypericum*. Recently, a veterinary oral surgeon called to inquire about postoperative homeopathic treatments for his patients. Bleeding had been reduced with the laser, but postoperative pain remained a concern. Dosing with *Arnica montana* 30C and rinsing the tissues with *Hypericum* and isotonic saline solution (five drops per ounce) proved to be the solution. When he called expressing his appreciation, he stated that the animal was able to eat the same day as surgery.

Both man and beast may suffer from toothache, gum pain and extensive dental work. Significant relief may be obtained from using a *Hypericum* saline oral rinse. This rinse may be prepared by adding five drops of the mother tincture to one ounce of isotonic (0.9 percent) saline. This solution may be applied to wounds anywhere on the body. It is safe enough to be used as an eyewash.

Osteoarthritis

Joint inflammation is common in older animals and often the cartilage and bone is involved. Roughened joint surfaces may be easily visualized on radiographs and are sometimes detected by manual palpation. Cattle on concrete suffer more than those on pasture. Rubber mats in tie stalls and well bedded box stalls are humane housing for confined animals. Dry bedding or clean sand makes a good surface for reclining and rest. The following medicines will bring relieve in those animals suffering from the pain and stiffness of osteoarthritis:

Rhus toxicodendron is indicated when stiffness is worse in the morning and in cold, damp weather. Even though she experiences great pain on beginning to move, the longer she stays in motion, the better she feels.

Hecla lava is indicated when there are large bony deposits (exostosis) around joints near the hooves. Cattle needing *Hecla lava* may have a rather sudden drop in milk production. The bones of the head, face and jaw may also be affected, making this medicine an especially good therapy for Actinomycosis (lumpy jaw). Affected cattle hate to have pressure applied to bony lesions.

Ruta graveolens will relieve stiffness that is better from pressure. The bones feel bruised, but are better applied pressure or receiving a massage. Weakness in the lower back is under the sphere of this medicine. Dosing BID or TID is indicated when the hips, stifles, or carpal (wrist) joints are affected. *Calcarea phos* is a complementary medicine.

Calcarea fluorica is known as "bone salt" and relieves older patients with long-standing arthritic symptoms. These patients may also have enlarged, stony hard glands. They are worse from change of weather, cold drafts of air, and from re-injury to the joints. They are better from warm applications, rubbing and massage. Dosing is recommended according to the intensity of symptoms. For moderately stiff cases, daily for 14 to 28 days and for the truly chronic, dose once weekly for two to four months.

Ovaries and Infertility

Cimicifuga racemosa (black cohosh) is indicated for inflammation of the ovaries and pelvic canal (pelvic inflammatory disease). It is also used with success in cases where there is obvious pain on examination and a history of miscarriages in the first trimester. Anestrus is common with associated pain in neck and back.

For anestrus associated with thick, yellow, odorous discharges from the uterus, use *Hydrastis canadensis*. In these cases the ovaries tend to be small, hard and painful. Often there is jaundice with inflammation of the liver when this medicine is called for. The patient will be lean and tired with a worn out appearance.

An animal whose ovaries are shriveled, and that also has dry, weathered skin and rough coat, especially with pain over the right ovary, requires *Iodium*. Give twice daily for three days in proestrus and once daily for ten days in mid-cycle.

In animals that tend to have small ovaries along with high milk production, *China officinalis* may be indicated. For example, a cow calves in good body condition with moderate milk production. As the milk production increases, she begins to lose body fat. When a critical percentage of body fat is lost, the ovaries become inactive. In these cases, an increase in total digestive nutrients is strongly indicated along with homeopathic therapeutics. Give *China officinalis* daily for ten days.

Paralysis

See Calving Paralysis

Parasites

See Worms

Pink Eye

See New Forest Eye

Respiratory Affections

Early symptoms of fever, thirst and cough call for *Aconite*. Later stages respond to *Byronia alba* when the cough is dry and the patient has obvious chest pain. The patient calling for *Bryonia* often refuses to move and may be found standing very still or lying down. Cattle with a moist, productive cough are helped by *Antimonium tartaricum*. Fevers that persist indicate a need for *Ferrum phosphoricum* or *Sulphur*. Animals with chest tightness, laryngitis and increased thirst for cold drinks call for the homeopathic medicine *Phosphorus*. In all cases, the recommended dosing is TID for five days.

Scar Tissue (Cervical and Vaginal)

Prevention of cervical and vaginal scar tissue can be accomplished with the judicious use of *Calendula* topically and in potency. After a difficult birth when tissues have been contused and lacerated, *Calendula* 30C or 200C given twice daily will prevent much scarring and disfigurement of the vulva, vagina and cervix. *Calendula* lotion or ointment applied daily will help prevent infection. *Graphites* is also of use (dose twice daily for one week).

Stomatitis

Inflammation in the mouth can have several causes and concurrent symptoms. If at teething, *Calcarea phosphoricum* will aid with their eruption while reducing pain and swelling in the gums. When accompanied by greenish diarrhea in calves, *Chamomilla* covers the simillimum.

Complementary therapies are high-calcium minerals, probiotics and antioxidants such as Co-enzyme Q 10, beta carotene, vitamin C, B-complex and A, D and E.

In older cattle, stomatitis may be associated with bacteria, mycoplasma or viruses, and often associated with nutritional deficiencies. There may be strong offensive odors to the breath. The tongue is swollen and the gums tend to be spongy and bleed easily. When these symptoms arise, *Mercurius solubilis* will speed improvement. More severe symptoms in untreated cattle with drooling of offensive saliva, dysentery and straining to stool, calls for the prescription *Mercurius corrosivis.* Recommended dosing is BID or TID for seven days.

As with younger cattle, nutritional correction is in order with emphasis on the antioxidants, beta carotene and vitamins C, A, D, and E. Serology for presence of viruses in the herd may be in order if stomatitis persists in a herd.

Teat Injuries/Ailments

Use *Calendula* non-alcoholic spray topically four times daily or *Calendula-Hypericum* spray topically four times daily or *Calendula* ointment after each milking. Choose one depending on what you have available. In addition, give *Arnica* 12X or 30C orally four times per day for four days. The above treatment will aid in preventing mastitis and/or scar tissue from forming. For chapped teats combine nine ounces of olive oil and one once of *Calendula* mother tincture for a soothing oil; dissolving five pellets of *Hypericum* 30X in an ounce of water with subsequent succussion (agitation) becomes an energized addition to the *Calendula* and olive oil.

During the monthly herd examination, Mr. M. stated, "I just wanted to thank you for the recommendations for treating teat injuries. Since we have been giving *Arnica* orally and applying non-alcoholic *Calendula* to the teat, we hardly ever have a case of mastitis after a tramped teat."

Ticks, Lice and Insects

Ticks become active early in the season in eastern Pennsylvania. People begin to remove them from their pets, horses and themselves during the last week in March. As a result, the cases of

Lyme Disease are on the increase. There are many non-chemical alternatives which have worked well: Mountain Trail Buzz-Off, a combination herbal ointment that can be applied to the face, neck, ears and anyplace on the body where ticks, flees or mosquitoes are biting. *Staphysagria* 3X, a homeopathic dilution, can be added to the drinking water or applied as a spray or wipe on. A one-percent solution is safe for people, livestock and pets. This may be mixed with Shaklee's Basic H in a one-percent solution to decrease surface tension. The homeopathic dilution *Ledum* 6C or 6X, can be applied directly to the skin where ticks or mosquitoes or other insects have stung the patient. Garlic pearls or garlic oil, when added to food daily, acts as repellant. Free choice foods such as cold water kelp, diatomaceous earth, real salt and humates should be offered to all species of livestock. (See also Bites and Stings.)

Ulcers (Stomach or Intestinal)

High-stress management with crowding and acid-producing feeds are the primary factors in ulceration of the GI tract. Chronic grain overload of feedlot cattle will produce clinical disease, as will heavy metal poisoning. Clinical conditions that accompany or follow ulceration are rumen acidosis, laminitis, liver abscess and various forms of hepatitis. Ulcers associated with Hardware Disease (Traumatic Reticuloperitonitis) and Vagal Indigestion are common in some herds at certain times of the year.

Changing of diet and management will do much to encourage sick cattle to survive. It is noteworthy that animals on a rotational grazing system with lots of space rarely develop ulcers. Feeding oats rather than corn, wheat or rye, along with good quality grass hay, will be supportive. Avoiding all high-nitrogen feeds like soy, urea and synthetic protein is strongly recommended. The following homeopathic medicines will speed the healing process and reduce suffering:

Arnica montana 30C or 200C. At the first indication of pain, the cow's head may be extended and each breath or movement makes the pain worse. She prefers to lie quietly without much pressure on the chest. Recommended dosing is three or four times daily for three days.

Carbo vegetabilis 30C. Bloating accompanies Vagal In-digestion. The patient has a cold muzzle and breath yet prefers fresh air. The entire venous circulation is compromised (see indigestion) and the animal tends to collapse in a warm, stuffy barn. She should be provided with a fan or moved to a well-ventilated area. Dose hourly for up to seven doses, or until bloat is relieved.

Phosphorus, 30C, 200C, or IM. This is a good follow-up medication to the first two because of its specific effect on tissues. The patient is thirsty, may have a cough and tends to be restless. They may feel better from rest and sleep, but rarely lie on the painful side. If it is a stomach ulcer, she lies on the right side; whereas, with a liver disorder, she will lie on the left. Recommended dosing is TID For five to seven days.

Ferrum phosphoricum, 30C and *Ferrum metallicum* 200C. Bleeding ulcers lead to weakness and anemia. Once the patient is stabilized, even though pale, these two medications may be alternated to stem the blood loss and conserve hemoglobin. Recommended dosing is once or twice daily for up to fourteen days.

Urinary Tract Affections (Nephritis)

Kidney and urinary tract health are necessary for cattle to be productive. Most often urinary tract infections are secondary to reproductive tract infections. Bacteria and other pathogens are known to ascend from uterus and vagina via the urethra into the urinary bladder. These affections are more common in the female than in the male, and more probable in hot, humid weather when the disease can take a rapid progression.

At the present time Leptospirosis, called Bright's disease in homeopathic literature, is on the increase in all species. Once the diagnosis is made, appropriate antimicrobial therapy is indicated. The following homeopathic medicines will support cattle in recovery:

Cantharais has a picture of straining with bloody urine. The pains are burning or cutting and can occur anywhere along the urinary tract from either kidney all the way down to the urethral meatus. The irritation (pain) attends the act of urination and is present afterwards. The patient is violently worse during these times. She may be angry and restless, but not particularly thirsty. Dosing three or four times daily for three days is recommended.

Phosphorus follows the acute prescription well in urinary disease. In addition to blood in the urine, the patient is fearful and thirsty. She likes the water fresh, cold and abundant. She urinates often and copiously. Concurrent disease of the liver with accompanying jaundice calls for this medication. Dosing three times for five to ten days is recommended. Two or more potencies of *Phosphorus* may be needed until the patient is cured.

Natrum muriaticum is compatible with *Phosphorus* and follows it well. In general, the patient has a high salt craving to go with excessive thirst. There is dryness to the mouth and the membranes. She is worse from hot, sunny weather, frequently seeking shade and drinks. Renal pain is worse after urination. Dosing with 200C or higher three times daily for five days should bring relief when symptoms match.

Apis mellifica has burning pains extending to the bladder with little thirst, but frequent need to urinate. The urine passes one drop at a time. Apis is strongly indicated in edema and puffy swellings secondary to kidney disease. Other keynote symptoms are fresh cow mastitis with udder edema, and allergic reactions to insect stings. Symptoms are better from cold baths, shade and cool air. The patient is worse in hot weather, form the hot sun, and from warm, stuffy barns (like *Pulsatilla*.) Recommended dosing is TID for three days.

Berberis vulgaris has inflammation of the kidneys with red brick dust sediment in the urinary sediment. The pain radiates from the kidney region to other parts of the body, for example, the lumbar muscles, the hind legs or the bladder. Pain occurs in the thighs and loins when urinating. There may be accompanying radiating pain in the region of the gall bladder or liver. The patient is better lying in a restful position. She is worse from jarring, moving, standing and urinating. Recommended dosing is TID for five to seven days.

Warts

Prevention is the key to this problem, especially Hairy Wart Disease and Strawberry Wart Disease. Nutrition and hygiene are invaluable ingredients of any good program. Remove all animal proteins, fats and blood meal from your feeds. Continue to daily provide kelp and minerals from natural sources. Eliminate the use of such hormones as BST, prostaglandins and steroids. If you vaccinate, use the fewest antigens at one time. For livestock that are confined, provide fresh, dry bedding every day. Clean hooves daily or use a foot bath at least once weekly. Pastures need to be mowed, rested and treated with compost or other high carbon material. When outbreaks occur in a herd, your choices of treatment may include:

Kreosotum 200C. Lesions are continuously moist and have an offensive odor. Erosions of the coronary band are painful and lead to poor hoof growth. Concurrent bandaging with *Graphites* ointment may promote healing. Dose once daily for seven days.

Nitric acid 30C. Coronary band is very sensitive to pressure and touch. Lesions tend to be jagged and resemble warts with deforming scars of surrounding horn. There may be concurrent ulceration in tissues of the mouth or rectum. Dose twice daily for five days. Concurrent bandaging with *Hypericum* ointment is a helpful adjunct therapy.

Thuja occidentalis will be used in cases with deformity of hooves and presence of fibromas or warts. The use of vaccines appears to cause aggravation of symptoms. If you are in the middle of a hairy wart disease outbreak, postpone vaccination until the animals have recovered. A nosode has been prepared from affected tissue and is available through licensed veterinarians for use in prevention of this disease. *Thuja* 30C may be given daily for a week. The nosode may be given every 14 days for up to three months. *Thuja* ointment under a bandage will aid in removing horny tissue.

Worms

The question of worming in the domesticated animal by homeopathy is a contentious one inasmuch as homeopathic remedies do not actually act as vermicides. The theory behind homeopathic worming is based on the belief that these remedies, while not actually killing worms, will render the stomach and intestinal tract unsuitable for the establishment or development of worms. This has proved successful in and cattle and other animals.

The presence of worms in general is denoted by staring; hidebound coat; appetite at one time poor, at another greedy; loss of flesh and condition; and occasional attacks of colic, or diarrhea and dry cough. The farmer may elect to use a conventional wormer and then follow with the appropriate homeopathic remedy, or they may choose to use the homeopathic medicine alone. Fecal examinations before treatment will assist the veterinarian and the farmer in determining the specific parasite species and in matching it to the specific homeopathic medicine. For Round-worms (Ascaridae) one might choose *Cina, Abrotanum,* or *Santoninum.* Tapeworms (Taenia) may require *Granatum, Kamala, Filix mas,* or *Chenopodium.*

Follow the homeopathic treatment with a fecal count in one month. Repeat homeopathic treatment if necessary. One farmer used to worm every eight weeks; he can now stretch it to every 12 to 14 weeks. If needed, an invasive wormer such as pyrantel or benzimidazoles may be utilized in alternation with the homeopathic regimen.

Historically, the 3X and 6X potencies are dispensed by prescription only and thus communication with your local veterinarian is imperative. The homeopathic medicines and dosages given in the chapter on Keynotes are suggestions only. Each herd situation will be different. These medications are best given as part of the total holistic health care program. As with all other medications, care should be taken to store them away from the inquisitive eyes and hands of children.

Wounds/Injuries

This calls for *Arnica montana* 12X or 30C every one-half hour for three doses. If the *Arnica* does not control the bleeding, *Phosphorus* 30C every half hour may be given. If the animal indicates that the wound is painful, *Hypericum* should then be considered at a dose of 30C or 200C TID for three days.

8.

The *Materia Medica,* Keynotes & Nosodes

While a student at the Penn State University main campus in 1965, 1 became fascinated with the library. There were volumes that I had never found at Gettysburg or any of the other colleges in central Pennsylvania. My future bride was over 100 miles away completing a tour of duty at a hospital in another state. Hence I had time to spend in the library.

Library research was not a new avocation for me. As a farm boy, any available time was devoted to reading. Hours recuperating from various on farm injuries were filled with the reading of books and magazines on various subjects. Veterinary medical texts were of special interest. Penn State library provided a myriad of old veterinary journals. On one occasion, I stumbled upon two older bound texts that I had never seen. The titles at first meant nothing to me. I had to think about the titles and explore the pages because the names therein were foreign—completely unfamiliar. The titles of those two old texts were *Materia Medica* and *Veterinary Materia Medica.* I had not a clue what I was looking at and returned them to the shelf after a brief review. That seed of information laid dormant in my psyche until that initial homeopathic course at Millersville University.

The term *materia medica* refers to those texts that contain the listing of medical materials that a medical practitioner will incorporating into his practice. Julian Winston in his text *The Heritage of Homeopathic Literature* describes the homeopathic *materia medica* as being, "Quite different from that of its allopathic counterpart, in that it is not only a listing of the chemical makeup of the material and its possible uses, but, most importantly, it is a detailed listing of the symptomatic response of a healthy person under controlled circumstances [proving] to the medicinal influence." Each *materia medica* that was written after Hahnemann's work (with the exception of other provings) is essentially a record of personal experiences with the remedies that Hahnemann outlined or remedies that were added through new provings.

These types of texts are at the core of any serious medical discipline. Every physician of every age had a *materia medica*. Hippocrates used a *materia medica* in his practice. Hahnemann while translating Cullen's *Materia Medica* from English to German questioned the author's references to Cinchona bark being the cure of Malaria because of its bitter taste. That query caused Dr. Hahnemann to travel down the discovery road to homeopathic principles.

The Medicines

Abrotanum (Southernwood)

Clinically, this remedy has been helpful in the control of roundworms (Ascaridae) and oxyuris. The southernwood plant grows on sunny hills in southern Europe and is cultivated in gardens. It is thought to destroys worms. Dosage: 3X or 6X potencies given daily for 14 days.

Aconitum napellus (Monkshood)

The common name is monkshood, but is also called *Aconite*. The active principle is a potent alkaloid. Keynote symptoms are: fever, sudden fever after stress, chill or shipping, grooming (a haircut). Accompanying the fever there is fearfulness, restlessness and increased thirst. The pulse is hard and accelerated, and respiration may be labored. It is best to give *Aconite* early in the course of disease. Use this medicine first for eye injuries and infections, shipping diseases, laminitis (acute founder) and acute mastitis. The harsh dry cough is a predominant symptom. Symptoms are generally worse at night. *Aconite* should be strongly considered for the early stages of any respiratory disease. Other possible characteristics are a hot, dry muzzle; hard, dry, insufficient evacuations; irregular or suspended rumination. Administer *Aconite* 12C, 30X, 30C or 200C three or four times daily for three days immediately upon arrival to the farm. *Aconite* has shown itself to be helpful in preventing illnesses brought on by the stress of shipping.

Antimonium tartaricum

The keynotes are moist cough with no expectoration. The debilitated patient has drowsiness and is not able to clear the airway because the cough is so weak. The rattling respiration is characteristic and can be detected without a stethoscope. General dosing is TID for up to five days.

Apis mellifica (Honeybee)

Indicated for swellings (internal and external dropsy), udder edema, hard quarter mastitis, right-sided ovarian cysts, and when manure is watery brown. Affected areas are worse from warmth and external heat and better from washing or moistening affected area with cold water. Cattle will seek shade, but will not drink even though they need to be rehydrated. One or two doses often will remind them to drink. General dosing is TID for three days. For reaction to insect stings, dose every one-half hour to affect.

Arnica montana

The medicine of first choice in any case of trauma of any kind. Injuries to soft tissues, worse to touch, trauma to body and mind. Use first in bruising, bleeding and disorders of muscle and the heart. Concussions, contusions, hematomas, nose bleeds. Oral and topical preparations are available. Better from rest and lying down. Worse from work, touch, motion. General dosing: in cases of bleeding, hourly for three doses; in cases of trauma, three times daily for three days. Potencies of 30C and 200C are available in the homeopathic farm kit. With these potencies, withdraw meat zero days, milk zero days.

Arsenicum album

The crude mineral arsenic trioxide is associated with nausea, toxic indigestion and hemorrhagic enteritis. Keynote symptoms are: acrid or burning diarrhea and loose, offensive stools; fever and restlessness; thirst for frequent small drinks; and chilliness of body extremities. *Arsenicum album* is useful for many septic and toxic conditions. Past drugs and vaccinations may have lead to chronic disease, which may respond to *Arsenicum*. Useful as patient attempts to excrete poisonous metabolites. Nightly aggravations and exhaustion from slight exertions. All discharges tend to be burning and irritate the skin around the body orifices (nose, lip eyelids, perineum). Cattle will often be lying down from exhaustion. Stool tends to run from the rectum with only mild tenesmus (straining). Calves benefit from the curative action of *Arsenicum album* during white scours. Offering electrolyte solutions between feedings or instead of milk feedings is supportive, as are probiotics three times daily. Most are thirsty and will drink water or electrolytes frequently. Dairy animals with a septic mastitis and an offensive, watery diarrhea call for the use of *Arsenicum album*. Follow-up medicines are *Phosphorus, Lachesis,* and *Sulphur.* General dosing is TID for three to five days.

Baptisia tinctoria

Wild Indigo affects the body like as severe case of influenza. Fever occurs, which makes the patient feel toxic or drugged as though in a stupor. A chill at 11 a.m. gives rise to a high fever in the evening. The tongue is coated yellow with brownish streaks and all secretions and discharges are offensive. Useful where there has been poisoning from food or water and where dysenteric stools have a putrid odor. They are worse from cold wind, but also from hot, humid weather. (Entry written by Michael Reece, ND.)

Belladonna (Deadly Nightshade)

Belladonna is indicated by sudden febrile attacks with delirium and threatened convulsions. Dose every one-half hour for two or three doses with 200C to effect relief. Medicines that follow well are *Bryonia, Calcarea carbonica, Phosphorus* and *Sulphur.* Other characteristics are excessive increase heat of the head, muzzle (with dryness), roots of the horns and glands; quick pounding pulse; hurried, panting breathing; hard, dry and almost black excrement; bloody urination; swollen glands in throat. Eyes are bright and inflamed with dilated pupils. Intense thirst coupled with aversion to drink. Uncertain and faltering step; the tail is arched and carried high, and constantly lashing the flanks.

Belladonna and Mastitis: A Case Study

At morning milking, a Jersey cow showed signs of mastitis—a high temperature of 108 degrees and a hot, hard quarter thick with pus. The mastitis also involved a supernumerary teat. Normally these "fifth teats" are removed before the animal is bred. Lacking the time to treat the animal frequently, the owner chose to only treat morning and evening with *Belladonna* 200C. No change was observed after 12 hours. That same evening, the owner decided to become more aggressive with treatment and began dosing the cow every 15 minutes with *Belladonna* 200C along with frequent stripping of the quarter. "During the next two hours, I watched that temperature come down and that fifth teat dry up," stated the farmer. "I was amazed." The first dose was given at 8:00 p.m. By 10:00 p.m., the temperature was 104 degrees followed by a morning reading of 102 degrees. The appetite returned to normal within 24 hours.

Bellis perennis (English Daisy)

Bellis perennis is useful when there is repeated bruising. Follows *Arnica* well and helps prevent fibroids and tumors at locations of bruising.

Berberis vulgaris

Renal and biliary disorders with a tendency to form calculi call for *Berberis vulgaris*. Predominant symptoms are kidney stones, gall stones and brick red sediment in the urine. Painful conditions radiate to other organs of the body, for example, from the kidneys to the lumbar muscles, or from the kidneys to the bladder, or from the liver to the kidneys. Often cattle have pain in the lumbar muscles. The patient is worse from motion and from standing, and better from lying down.

Bryonia alba

The common name for the plant is white hops. All symptoms are made worse from motion. The sick patient spends much time standing or lying, rarely moving. Keynote symptoms are: dryness of throat and all membranes with thirst for large quantities of water. There frequently is a hard dry cough while the patient tries to be as still as possible. In cows it is indicated when there is painful swelling of udder that is aggravated by motion.

Calcarea carbonica

Carbonate of lime from the middle layer of the oyster shell has affinity for skeletal and lactiferous tissue in the body. Large, blocky, heavy cows respond well to this homeopathic preparation, especially many in the Holstein breed. This cow is a heavy milker that maintains her body condition even in mid- or late lactation. Cattle prone to milk fever, foot disorders and swollen joints (hocks) call for this prescription. Dose frequently in down cows that do not try to stand after intravenous calcium therapy. Other keynote symptoms are flat warts in heifers, tendency to obesity and illness after change of weather. Patients calling for *Calc carb* are better from dry weather and worse from cold, wet weather and from becoming chilled.

Calcarea fluorica

Useful in the prevention of osteoarthritis of the coffin joint and recurrence of corns. Used frequently in cases of bone spavin, this medicine will also promote stronger hoof growth. These cattle are better from warm weather and worse from weather changes.

Calcarea phosphorica (Phosphate of Lime)

Calc phos is indicated in dairy-type cows (lean with rigid fiber and look like a wedge of cheese if viewed from above); young growing animals; active cows; those in lean body condition who do not like restraint or being examined; strong appetite, aggressive within herd; calves tend to be susceptible to digestive upsets; mature cows tend to get decreased rumen movement, diarrhea or constipation. *Calc phos* is given in dry period to help prevent milk fever; symptoms are better from warm dry areas but worse from changing, cold, damp weather. Used to treat lameness; rheumatic affections in the joints, exostosis, caris of the bones, carious ulcers. Also indicated for frequent painful and/or difficult emission of large quantities of urine; dark, sometimes burning urine. In females, used for animals that have episodes of anestrus, ovaries develop slowly, tend to be inactive and may be stuck in the luteal phase of the cycle.

Calcarea phosphorica is important in the development of bones, joints and teeth. For prevention of stunted growth dose with 12C, 30C or 30X weekly. For treatment of limb or dental deformities, dose daily with 12C or 30C for two weeks.

Calcarea phosphorica: A Case Study

Holly, a Holstein, calved in July. She was milking well with normal reproductive tract, but overall had lost weight following calving. She was prescribed *Calc phos* 30X once daily for one week and improved. *Calc phos* has an affinity for tissues as related to the growth and repair of cells.

Calendula

Made from marigold flowers and leaves, *Calendula* is one of the most useful remedies for any sort of ulceration or open wound. It can be obtained in many forms including pellets, ointment, lotion and spray. It is reliable and will bring about healing of tissue as well as promote healthy granulation. Useful in treating wounds of the eyes, especially those from blunt trauma. It is often used with *Hypericum* when treating open wounds where there is also pain. It is used for calving injuries, topically to treat cracked teats, and as an infusion in cases of metritis (uterine infections).

Cantharis (Spanish Fly)

Helpful in burns (rawness and smarting, relieved by cold application), sunburn, stings (swollen and inflamed, burning and biting pains), bladder irritation with frequent scalding, painful urination, digestive disturbances with burning sensations in esophagus and stomach. Condition is better from gentle massage, and worse from touch, and urinating.

Carbo vegetabilis

Carbo veg has the reputation of "bringing them back from the dead" because of its work over the past two hundred years in shock, fainting and frostbite. Patient is cold but desires fresh air. Other keynotes are bloating with abdominal distention and indigestion within hours after overeating rich grain or forage, better with belching. The modalities are worse in damp weather and better from belching and being fanned. *Carbo vegetabilis* has some symptoms similar to *Arsenicum album* The patient has coldness of breath, skin and extremities. Fainting and collapse may occur from distention or from being in warm stuffy stables. Fanning the nostrils often brings relief. As with *Arsenicum album* cases, stools are usually burning and offensive.

Carbo vegetabilis: A Case Report

A herd of 40 Holsteins were grazing on fresh alfalfa pasture most of the day. When the farmer brought the cattle into the barn for milking the following symptoms were observed: bloat and distention, shortness of breath, loss of appetite and staggering gait. The alert, experienced dairyman knew the prescription. He treated each affected cow with *Carbo vegetabilis*. One half hour later each patient was given a dose of *Nux vomica*. All but three cows responded immediately. These three were given a second dose of *Carbo vegetabilis* and went on to complete recovery.

Caulophyllum (Blue Cohosh)

Used to prepare uterus for labor and delivery and for labors that do not progress (interrupted contractions with failure of cervix to dilate). Higher potencies stimulate a "worn out" uterus. *Sepia* may also be utilized when dilatation of cervix has begun but labor is not progressing. Dose with *Caulophyllum* three times daily after delivery to help expel placenta. Other indications are rheumatism of small joints with erratic, shifting lameness.

Causticum

Used primarily for progressing weakness of muscles, for tremors, contracting tendons and paralysis. Can also be used for skin problems such as tenderness in skin folds or between thighs or warts that bleed. *Causticum* is used in dairy cattle when there is retention of urine with involuntary passage of urine when walking. In this condition, general dosage is *Causticum* 30X or 30C once daily for 10 days. Of the 116 medicines listed in *Kent's Repertory* for retention of urine and involuntary passage of urine when walking, four are commonly used in the dairy cow—*Silicea, Gelsemi-um, Sepia* and *Causticum*.

Chamomilla (German Chamomile)

Indicated for extreme restlessness, moving from one place to another; lying down and immediately rising again. Diarrhea; evacuations mingled with stringy or filmy white substances. Roughness and harshness of the skin and coat. The muzzle may be heated and dry; tongue is fissured. Animal is fretful, irritable and unmanageable with diminished appetite and eagerness to swallow cold drinks. Emission of flatulence, upwards and downwards. Viscid, stringy appearance of the milk; teats and udder are inflamed; discharge of milk is suppressed. Dry cough, occasioned or aggravated by the pressure of the throat; hoarse and noisy respiration. Crack-ing in the joints; development of tumors (swellings) about the joints of the legs. In young animals, *Chamomilla (Cham)* is indicated by colic or diarrhea during teething. Stool looks like chopped spinach; gums are swollen, red and painful; patient refuses warm milk, prefers cool drinks. Dose four times daily with 12X, 30X or 30C. *Chamomilla* is compatible with *Belladonna* and *Pulsatilla*. The baton may be passed from *Belladonna* in an acute febrile condition to *Chamomilla* when symptoms change to just diarrhea and gingivitis. A good remedy for young livestock.

Chelidonium majus

A prominent liver remedy with general lethargy and indisposition; tongue coated dirty yellow; mucus membranes with signs of jaundice, and clay-colored stools which may alternate from diarrhea to constipation. The symptoms of the animal will be worse on the right side, with motion, touch, change of weather and very early in the morning. Past exposure to hormones, herbicides and other xenoestrogenic substances cause livestock to become ketotic when under stress.

Chenopodium (Jerusalem Oak or Wormseed)

In addition to controlling tapeworms, it has been helpful in cases of strongyles and hookworms. This rank, odorous plant grows about waste places in almost all parts of the United States. The seeds of this species are in common use as a vermifuge or an agent that expels worms. Dosage: 3X or 6X potencies are helpful given daily for 14 days.

China officinalis (Cinchona officinalis)

Also known as Peruvian Bark, this medicine is indicated for loss of vital fluids (milk, blood, perspiration, or from diarrhea) leading to a lean or thin, dehydrated state; depressed metabolism; gas with no relief from passing it; weakness, nervousness and debility. Females tend to have small ovaries along with high milk production. Particular symptoms include undigested frothy stools with much gas and rumbling, along with pain which is better from hard pressure. The weakness follows fluid loss in chronic cases. All symptoms come and go in periodic fashion—for example, the same time each day or every 14 days.

Cicuta virosa

Water Hemlock poisoning affects the central nervous system as well as the entire length of the GI tract. Keynote symptoms are bending of the head, neck and spine backwards, or having the head twisted to one side. The patient may have convulsions with colic and the tendency to fall to one side. Often the jaws are clenched and there is grinding of teeth. They are worse from touch, jarring and any sudden stimuli.

Cimicifuga racemosa (Black Cohosh)

Used for pelvic inflammatory disease (inflammation of the ovaries and pelvic canal). Anestrus is common with associated pain in neck and back. The patient will experience pain from the internal examination. There is history of miscarriage in first trimester. Useful for muscular and neurorganogenic pains of neck, back and pelvis. Large animals become tense when sensitive organs are palpated during a veterinary exam.

Cina (Worm Seed)

This drug does not consist of seed, but of underdeveloped flowers mixed with the scales of the calyx and pedicel of different species of the genus *Artemisia*. Animals needing *Cina* tend to be irritable and nervous and have diarrhea with discharge of roundworms. According to Christopher Day, MRCVS, they may show "signs of colic with tooth-grinding, have a distended and hard abdomen, and the dung often contains white mucus." Dosage: 3X daily in the drinking water for 14 days.

Colchicum autumnale

Gastrointestinal inflammation with trapped gas and bloat are keynote symptoms. Ruminants like cattle are distended on both sides of the body. The smell of food causes nausea and loss of appetite. There is a tendency to collapse in acute conditions. Many of these patients exhibit joint stiffness, especially with gout-like symptoms of the lower extremities. They are worse from movement (like *Bryonia*), in the autumn, and from sunset to sunrise.

Colocynthis

The keynote symptoms are abdominal cramps accompanying GI upset with the patient bending double seeking relief. Pain is ameliorated when the legs are pulled forward or one leg is drawn up under trunk. The patient looks to one side each time a spasm or cramp flies into his abdomen. They are better from pressure, warmth and bending double and worse from anger and lying on the painless side.

Conium maculata

Toxic effects of ingestion are ascending paralysis and motor nerve paresis, at first observed in the distal hind limbs. This medicine is strongly indicated after bruising or shock to the spinal cord. Eye symptoms are paralysis of ocular muscles, photophobia, conjunctivitis and keratitis (cornea turns hazy blue). Hard tumors may occur in the head, the breast, or at any location in the body. Like the action of *Bellis perennis, Conium maculata* will aid in the prevention of a tumor at the same location where the breast or perineum had been bruised. Older patients with colon or bladder weakness may benefit. Patients are better from moving about and worse from taking cold and from bodily injury.

Crotalus horridus

Keynotes are sepsis with hemorrhages of dark blood and yellow discoloration of the skin or mucous membranes. The blood refuses to clot. Lymphatic glands and vessels become swollen; veins may enlarge or burst. Illness may progress to a "purpurea haemorrhagica." There is postpartum bleeding like *Secale cornutum* with dark fluid blood that will not clot. Patients are better from being in open air and from rest; worse on the right side and in the damp, warm weather of springtime.

Cuprum metallicum

Keynote symptoms are spasmodic cramps that begin in the feet and ascend up the limbs to the body. The stiffness and cramping of the leg muscles spreads upward to the chest and abdomen. There may be colic with greenish diarrhea like *Chamomilla,* and the convulsions may be either tonic or clonic. The patient is better from drinking cold water and worse from touch, pressure and being startled.

Drosera rotundifolia

All respiratory organs are under this plant's influence. Medium or long-standing pleuritis (pleurisy) with dry cough call for its use. Some patients cough and cough until they vomit as is witnessed in human whooping cough. Virus pneumonia in calves responds to this remedy after *Ferrum phos* or *Aconite*. Primary laryngitis usually responds to *Phosphorus* or *Drosera*. Like *Silicea, Thios* and *Graphites,*

Drosera is useful in dissolving scar tissue. Patients are better from pressure and in open air, and are worse from warm drinks and in warm barns.

Echinacea—Rudbeckia

Indicated in cases of acute sepsis, boils and Puerperal fever. It is often combined or alternated with other therapeutics. For example, in gangrene alternate with *Lachesis;* in mastitis and metritis alternate with *Pyrogenium.* In cases of infection, dose with ascorbates (vitamin C) and probiotics.

Euphrasia officinalis

For thousands of years Eye Bright has been prescribed for various eye conditions. Its reputation as an effective therapy for conjunctivitis and corneal ulceration has been confirmed by homeopaths again and again over the last 200 years. New Forest Eye affections that result in much tearing so that cattle have a persistently wet face call for this remedy. Affected cattle have profuse acrid lacrimation so that the eyes burn, but the nasal discharge (coryza) is bland. These patients have photophobia and are worse in the light and better in the dark and from an opaque eye patch. Diluted tincture (10 percent in isotonic saline) can be applied topically, and the 30C pellets given orally.

Ferrum phosphoricum

The crude mineral iron phosphate, a tissue salt, is associated with feverish conditions. Keynotes are: fevers of 102-104, moderate speed of onset of symptoms, anemia with pale mucous membranes which flush with excitement or exercise, and upper respiratory infections with epistaxis (nose bleeds). Respiratory affections with moderate fevers in young livestock are within the sphere of the action of this remedy. *Ferrum phos* is useful for both bacterial and viral infections. Consider *Ferrum phos* in animals whenever the symptoms involve the throat or ears. A red, warm ear with an accompanying low fever calls for this prescription.

Filix mas (Male Fern)

The male fern is widely distributed in temperate regions and is found in abundance in most countries in the northern hemisphere, except in the eastern United States, where it grows somewhat sparse-ly in shady pine woods. The fresh root is gathered in July or August. Male fern has been long recognized as a vermifuge, especially for tapeworms. Dosage: twice a day for five days. The potencies 3X and 6X may be utilized.

Gelsemium

The Yellow Jasmine flower's medicinal picture is that of a tired old Basset Hound. Keynote symptoms are: dullness, drowsiness; muscular weakness and lack of thirst. Gelsemium is useful in many viral conditions, and in homeopathic prevention of milk fever and grass tetany.

Granatum (Pomegranate)

Pomegranate is a well-known vermifuge, especially for the expulsion of tapeworms (taenia). According to *Boericke's Materia Medica*, *Granatum* can be used as a vermifuge for the expulsion of tapeworm and homeopathically for the following symptomatic indications: salivation with nausea and vertigo. Dosage: 3X daily in the drinking water for 14 days.

Graphites (Black Lead)

Tissues are constantly moist with sticky discharges. Eruptions and discharges occur where skin folds, especially between the hooves. When a wound heals, it tends to from a keloid or cheloid—a fibrous tissue formation growing from a scar (in veterinary medicine this is referred to as an exuberant granulation or proud flesh). Indicated in animals with broad, heavy cervix with moderate fibrosis and associated eruption at the base of the udder or sores on the teat ends. Patient tends to be heavier and slower moving than normal and may suffer from bouts of constipation. The affected area and the udder can become heavy with weakened and stretched suspensory ligaments. Cracked nipples with a sticky discharge respond to local applications of *Graphites* ointment.

Gunpowder

Black gunpowder is a mixture, not a chemical compound. It has been useful for the prevention or treatment of various septic conditions. Keynotes for this homeopathic medicine are blood poisoning and infected wounds. In England, shepherds who had to care for sheep with foot rot would sprinkle gun powder on the bread and cheese to prevent themselves from contracting the infection from the sheep. Homeopathic gunpowder may be prescribed from the 3X potency up to 200C or higher. It appears to be effective in many potencies.

Hecla lava

Volcanic ash is the crude substance for trituration. This medicine shows affinity for the bones of the upper jaw and the lower legs. It has been useful in osteoarthritis and the very stubborn 'Actinomycosis' when gram-negative organisms with the same name invade bone and undermine the integrity of the osseous upper and/or the lower jaw. The affected parts are worse from touch and pressure. *Hecla lava* has medicinal value in many potencies from 3X up to 200C.

Hecla lava: Discovery and Uses

In his lectures George Macleod, MRCVS, would often share the interesting story of Mt. Hecla in Iceland. J. Garth Wilkinson of London was traveling in Iceland in the year 1872 when he noticed peculiar bony swellings on the sheep grazing some distance from the volcano that had erupted a few years earlier. The bony growths often became so large on the jaws of some sheep to make chewing impossible. Those poor suffering creatures needed to be slaughtered. Lactating cows and sheep grazing on the finer ash some distance away from the mountain would dry up— losing the ability to lactate. Several young horses grazing the same pastures died from lumps on the jaw bones so large as to cause dislocation. Livestock close to the volcano that grazed in the coarser ash had no obvious pathology. Macleod used to say, "Any substance that is so toxic to cause such terrible disease could be harnessed homeopathically to do much good, i.e., to heal people and animals suffering with the same symptom picture. Wilkinson collected some of the finer ash and had it triturated with milk sugar by a homeopathic pharmacist.

Garth Wilkinson adds, "I have used (the homeopathic) *Hecla* with excellent effect in toothache, gum abscess, swellings about the jaw and in difficult dentition. Clinical experience has shown the power of *Hecla* to arrest many forms of bone disease including osteosarcoma, scrofulous and syphilitic osteitis and exostosis."

The author has had several bone tumors of the jaw in llamas completely disappear, and a case of Bovine Actinomycosis go into remission for five years. A horse with osslets of the fore fetlocks responded to *Hecla lava* so that it can now trot and canter without lameness. The words of Garth Wilkinson and George Macleod still ring true in our ears.

Hellaborus niger

Keynotes are functional shutdown of parts of the brain due to high fever, head trauma, electric shock (lightning strike), or a stroke. The patient appears dull like *Gelsemium* and may even be comatose. Use in head trauma if another medicine is needed after *Arnica*. (*Natrum sulph* should also be considered.) Cattle and other farm species will hold fluid in the brain, pericardium, or chest cavity. Brisket edema is common due to threatened congestive heart failure. These animals are worse from electric shock, cold air, exertion, and from 4 to 8 p.m., and are better if kept warm. (Entry by Michael Reece, ND.)

Hepar sulphuris calcareum (Hepar sulph)

There is a tendency for all wounds to become infected; the patient is very sensitive to touch, pressure, chill and cold drafts of air; mastitis—overly sensitive to nursing, washing and milking; painful hoof infections and attending lameness; all pains better from warm soaks and wrapping up and worse from chill in any form. Helpful hint: with neglected painful wounds—choose *Hepar Sulph* and/or *Calendula*. Also useful in cases of Iritis.

Hydrastis canadensis (Goldenseal)

Used in anestrus associated with thick, yellow, odorous discharges from the uterus. Ovaries tend to be small, hard and painful. There is presence of jaundice with painful inflammation of the liver. The animal is lean with a worn out appearance, looks as though she needs an extended time of rest and recuperation. Herbal and homeopathically *Hydrastis* has a great reputation treating cancer of the liver and stomach. It cures many, but even in those not cured it relieves pain right to the end. [Grimmer in *Homeopathic Remedy Guide*, page 833.] This medication is strongly indicated in cattle with chronic digestive problems. Those cows that look old beyond their years should be given a three-month or longer dry period and receive *Hydrastis* once weekly throughout that time.

Hypericum perforatum (St. John's Wort)

Bruised nerves of lower legs and lower back, painful puncture wounds, injuries to toes, tails and teats, and weakness or paralysis. In the bovine, *Hypericum* is useful for obturator paralysis after difficult calving; puncture wounds from forks, wires and nails; painful teat injuries; and injuries to the tail. For the bovine species, dose 10-15 pellets for adults and 4-8 pellets for calves. If using dilutions, give comparable amounts.

Topical lotions can be applied to wounds for cleansing; followed by applying ointment to the clean wound. For eye injuries, use non-alcoholic lotion/spray, or prepare your own solution by adding 3-5 drops of mother tincture to sterile isotonic saline. Your veterinarian or physician can supply *Hypericum* mother tincture and isotonic saline. This solution may be applied to the eye as often as needed. *Hypericum* is the remedy of choice in photosensitization of non-pigmented skin in all species. Orally we recommend the higher potencies (200C or 1M) once daily while bathing the raw burns with *HyperCal* lotion several times daily as needed until healed.

Calving Paralysis and *Hypericum:* A Case Study

A farmer delivered a dead calf with a fetal extractor from a small Holstein heifer in May. Much bruising and tearing had occurred, which resulted in paralysis. Floating (swimming) in a tank on two separate occasions was unsuccessful. By the beginning of June, when he considered having her taken away, his son approached him with a new medicine, *Hypericum perforatum,* one that he was just learning about. After the first dose of *Hypericum* 30C, the appetite and thirst improved; after the second dose, she began flexing and shaking her legs. By the third dose, the heifer got on her feet and walked off. Seeing the need for diversity, this Pennsylvania farm family now uses rotational grazing for their 60-cow dairy herd. They also raise chemical-free birds and grow organic vegetables. Since the initiation into homeopathy both mammals and fowl have been placed on a holistic herd health program.

Ignatia amara

This medicine is wonderful at first calving and weaning. Animals calling for *Ignatia* have strong emotional ties to family members, mates and sometimes people as well. It is indicated in first-calf heifers that are tense and do not let down milk. At weaning, both mother and baby benefit. Digestive disorders and/or nervous behavior from grief or separation anxiety are common. There may be constipation, flatulence or loss of appetite like *Nux vomica.* Human patients needing *Ignatia* say that they cannot eat because it feels as though there is a ball in their throat or stomach. They are better from swallowing and from changing position; worse from grief, fright and unusual strong odors (tobacco smoke, coffee, other alkaloids).

Iodium (Iodine)

Shriveled, dry, weathered skin; rough hair coat; shriveled ovaries in the female and testicles in the male; pain over right ovary in female. Everything wastes away except the glands, which enlarge.

Restlessness and constant hunger are also present. There is usually evidence of thyroid gland imbalances. One keynote symptom is a ravenous appetite with corresponding loss of weight. The hair/coat tends to be rough and dry while the lymph glands become progressively more hard. Animals are often easily excited or hyperactive.

Ipecacuanha (Ipecac)

Ipecacuanha has a strong keynote of persistent nausea not relieved by vomiting. Another general symptom is bright red (arterial) bleeding from any body orifice. Particular symptoms are profuse salivation with a clean tongue, pain in the abdomen around the navel and loose stools that may be frothy or bloody or green. Bloody milk and coccidiosis in all species often respond to *Ipecacuanha*. Patients are better from open spaces and rest; worse in warm, damp barns.

Kali bichromicum

The keynotes are tough, stringy yellow mucous from body openings, especially the mouth, nose, sinuses and throat with loss of a sense of smell. There may be accompanied sluggishness, anemia and ocular inflammation with eyelid edema or corneal ulceration. Mastitis with stringy, watery milk calls for its use in lactating livestock.

Kali carbonicum

Keynotes are generalized weakness, sluggishness with intolerance of cold weather. There is a tendency toward obesity as well as thyroid disorders. In overweight cattle, it may be dosed in alternation with *Calcarea carbonicum* to help regulate metabolism. They are better from open air that is warm and worse from cold air in cold weather.

Kali iodatum (Hydroiodicum)

Lymphatic tissues of the throat and neck are strongly affected, tending to be swollen and hard. Glandular infections have a tendency toward sepsis or purpurea (see *Crotalus horridus*). Dis-charges are thick, greenish in color and offensive. Throat affections move toward the chest, just the opposite of true healing. Actinobacillosis and thyroid disease are within the sphere of influence of *Kali ioda-tum*. Macleod recommends it for Fog Fever and New Forest Eye disease. These cattle are better from cool, open air and worse from warm, moist weather. It is useful when calves are born in the late spring followed by long periods of hot, humid weather. Under those conditions, the medicine will prevent many cases with strangles-like symptoms and pneumonia.

Kali phosphorica

Kali phosphorica is one of the tissue salts that has influence over every tissue and organ of the body. Keynote physical symptoms are weakness and quivering of the muscles, along with weakness of the central nervous system and nerves. This may progress to paralysis of the back and leg muscles. Young, active animals with neuromuscular weakness that are worse at night call for this prescription. Older patients with threatened bronchitis or renal disease benefit. *Kali phos* gives great support to the circulation of the genitourinary system. They prefer lots of cold water and human patients ask for salty foods, sour foods and sweets. Perhaps these cattle will eat salted hay or silage before anything else. They are worse in the early morning and from physical exertion.

Kamala (Croton coccineus)

The plant grows in India, South Arabia and tropical Australia. Potencies are prepared from the powder obtained from the plant capsules and after trituration dissolved in alcohol. This is one of the main remedies to be considered in any worming program, especially tapeworm infestation. Dosage: dissolve ten powders of 30C potency in 500ml distilled water. From this stock solution 20ml should be added to the feed daily for four weeks. Ten powders would be equivalent to 25 #30 pellets.

Kreosotum

Beechwood creosote has been historically termed "flesh-preserver." Keynotes are offensive and excoriating discharges from the hooves or from any body opening. Violent, sudden desire to urinate, especially at night or while sleeping. In livestock, *Kreosotum* is useful for Foot Rot, Hairy Hoof Wart, and Thrush. Stomatitis with spongy, bleeding gums calls for the medication.

Lac caninum

The medication is prepared from whole milk of the dog, and has a profound effect on the mammary glands of many species. In lactating livestock, it is useful when mastitis moves from one side or one gland to another. The affected portion of the udder is tender and engorged with milk. Symptoms and pain are made worse from jarring, coughing and sneezing.

Lachesis muta

Heat intolerance, difficult swallowing, symptoms relieved by discharges. Problems begin in left-side; dark, watery blood; cellulitis; fever; purple tissues, septic mastitis, especially swollen left-side; septic metritis, left-sided ovarian cysts. Better in open air; worse from rest, sleep, pressure, warm liquid and in early morning. Intolerance to touching, or a constricting sensation around throat, neck, chest or waist. Gangrenous mastitis gives a dark brownish-red secretion instead of milk. The udder or breast will become hard and the nipples will turn blue or purple in color. The temperature of the udder may be either hot or cold. In cases such as coliform mastitis in cattle or sheep, animal improves with hourly milking. Lachesis patients are better from cold drinks, and worse from warm drinks and suppressed discharges.

Lacticum acidum (Lactic Acid)

Morning sickness, diabetes, and rheumatism offer a field for this remedy. Rheumatic pain in lower back, extending to shoulders, as well as rheumatic pain in wrists and knees with much weakness. Other symptoms include trembling of whole body while walking and pain in region of kidneys with frequent desire to urinate.

Lathyrus sativa

Paralytic affections of the hind legs, causing some muscle groups to be contracted. Human patients—calf muscles are so tight that heels cannot touch the floor. Muscle reflexes are increased. General neuromuscular weakness after any chronic illness calls for its use. They are better lying down and worse from cold, damp weather.

Ledum palustre

Puncture wounds better from cold; insect stings and animal bites; wounds feel cold and are better from cold; arthritic pains better from cold compresses and ice packs; skin eruptions without heat; and blunt trauma to eye (similar to *Symphytum*). In all species use *Ledum* for first aid after any injection reaction (vaccinations). *Ledum* may be applied topically and taken orally at the same time. Bites from rats, dogs and other animals and stings of spiders and insects that become itchy and cold indicate its use. Black and blue eyes respond very well. *Ledum* is much like *Hypericum* in puncture wounds. Often if you do not have a good response to one medicine, try the other. Ledum patients are better from ice water applications and joint pain is worse in warm seasons of the year.

Lycopodium clavatum (Club Moss)

Deranged digestion from poor function of liver and kidneys; bloat and gas after eating that is better with belching; patient feels full or satisfied on small portions of food only to be soon hungry again; fears loud noises and being alone, yet avoids crowds; physical symptoms tend to move from right to left. Within a short time after calving there may be indigestion or ketosis from poor liver function. Kidneys may be compromised, showing red sediment in the urine. Pockets of trapped gas are common and relieved by burping or passing flatus. The patient is better moving about and from warm foods and drinks. She is worse from 4 to 8 p.m., and in warm, crowded barns.

Magnesia carbonica

Calves not digesting milk well call for this medicine and that of *Calcarea carb*. The gastroenteritis leads to a sour body odor and poor growth so that they are thin and stunted. Heavy pregnant cows that are prone to milk fever benefit from *Mag carb* in conjunction with *Calc carb* and/or *Calc phos*. These may be dosed in combination during the dry period.

Magnesia phosphorica

Magnesia phosphorica is a polycrest that has many uses for man and beast. Keynote symptoms are neuralgic pains, muscular cramps, and abdominal cramps. All discomforts are relieved by warmth.

Modalities are extremely important when prescribing this remedy. Worse on the right side of the body and from cold in any form— cold weather, cold winds, cold washing and cold drinks. It is better from pressure, warmth, and bending double. Animals bend toward the painful side (like *Colocynthis*). Some veterinary conditions where *Mag phos* has been effective are milk fever, grass tetany, myosititis, and tying-up syndrome.

Mastoblast

A combination homeopathic oral product designed to boost the animal's own immune system to reduce or eliminate invading pathogens. It contains no chemicals which could injure the animal, nor is there any risk of objectionable residues in milk or meat.

Mercuris corrosivus or *Mercuris solubulis*

Both mercury salts have affinity for the digestive tract, the urinary tract and the central nervous system. Keynote symptoms are: ulceration of mouth or intestine with diarrhea and straining; thirst with profuse salivation; trembling of limbs; chills and fever with perspiration. Mercuris salts are useful in viral conditions and coccidia infections. *Mercurius corrosivus* has tenesmus (straining) not relieved by stool, or pain in the bladder not relieved by urination. All discharges tend to be dark and mucoid. The diarrhea observed with coccidia, winter dysentery and subacute BVD tend to match well with the keynotes of *Mercurius corrosivus*.

Mezereum (Spurge Olive)

Eczema and itching eruptions after vaccination. Regions of the body under its influence are the head, face, teeth, skin and long bones. The skin of the head or other areas often are covered with thick, leatherlike crust; ulcers appear with thick, yellowish white scabs, under which thick, yellow pus collects. Vesicles appear around the edges, itch violently, burn like fire; pain in periosteum of long bones; aggravated by cold air, cold washing, at night, by touch or motion; facial pain, warm compresses; toothache when teeth feel too long and decay at the roots (like *Thuja*). Ears and eyes become sensitive to cold drafts of air. Most useful in veterinary medicine for disease conditions that appear after vaccinations.

Murex purpurea

Prepared from the dried juice of the purple fish, this medicine has a profound action on the reproductive system. Keynote symptoms are ovarian pain with multiple cysts or multiple follicles in an over-drugged, worn out breeding animal. In cattle it becomes useful to help correct endocrine imbalances after prostaglandin and hormone injections.

Myristica sebifera

Often called the homeopathic knife. Like *Hepar sulph* the antiseptic action will hasten the draining of fluid and pus-filled pockets. Most frequently it is prescribed for affections on the sole of the foot or ears. All species benefit from this prescription when the site of illness is around toenails or fingernails. Usually it works well after prescriptions of *Hepar sulph* or *Silicea* have been tried and failed.

Myristica sebifera: A Case Study

Farmer Leroy called about a cow with a suspected abscess. "About five months ago I gave this cow an intravenous calcium treatment in the milk vein. She recovered from the milk fever, but about a month later we noticed a round swelling on one side in front of the udder. I think this is the same spot where I placed the needle. I had the local vet examine her. He said that there is a milk vein in that pocket of fluid somewhere, and he did not want to put a knife into it. I tried *Hepar sulph*, but nothing changed. It is now the size of a volleyball."

We sent Leroy two potencies of *Myristica,* a 30C and 200C. He dosed with the 30C first. On day three it drained a gallon of creamy yellow pus. Now that the body had opened its own pathway, Leroy could flush it twice daily with *Calendula* tincture and peroxide in saline. As of the last report, the cow had recovered. Five percent of cattle with abscesses do not respond well to *Hepar sulph*. Those animals call for the prescription *Myristica sebifera.*

Natrum carbonica

Like other cell salts, this medicine can have a strong influence upon the metabolism. Keynote symptoms are an inability to digest milk, poor digestion in general leading to gas and bloating, frequent diarrhea and weakness of the ankles. They are worse from the heat of the sun, from storms, from getting wet and from drinking milk.

Natrum muraticum (Sodium Chloride)

Most cows needing this remedy do not show strong estrus; ovaries tend to be large, some with fluid-filled pockets; cow tends to have puffiness in the legs and joints; worse from sun and hot weather, better from drinking cold water, shade and cool weather. All discharges are like the white of an egg. Animal may be prone to heat exhaustion. This remedy follows *Apis* well. *Natrum mur* patients crave salt, and may have weak kidneys. They are worse from hot weather and from 9 to 11 a.m. They are better from cool weather, cool rains and open air.

Natrum phosphorica

Natrum phosphorica has some symptoms like *Phosphorus* and some of the symptoms of *Mag phos* and *Kali phos*. With a lab name of "Phosphate of Soda," this tissue salt is found in most body organs and fluids.

Lactic acidosis and ailments from extreme acidity lead the prescriber to consider this medicine. *Natrum phos* is the tissue salt that is present when lactic acid is decomposed into CO_2 and H_2O. *Natrum phos* patients tend to have yellow discharges and have frequent acid indigestion. Acidosis with ketosis in ruminants is within the sphere of the action of *Natrum phos*. Animals with decreased kidney function may be given extended enjoyable life with judicious prescriptions of this remedy. *Natrum phos* patients are better in cool weather, and worse from high fat diets.

Natrum sulphuricum

One of Hahnemann's anti-sycotic medications, it helps the body to eliminate waste products. It is especially useful in young animals with stubborn conditions, or animals that have experienced head trauma. Like *Thuja occidentalis*, *Natrum sulph* is indicted for patients that have never been well since vaccinations. They are prone to develop bronchitis and painful chest conditions like *Bryonia,* and have diarrhea in the morning like *Sulphur.* These cattle become sick anytime that weather changes from warm and dry to cold and wet. They are better in dry weather and after eating breakfast.

Nitric acid

Painful conditions of the feet, around the mouth, rectum or genitals call for the use of homeopathic *Acidum nitricum.* The pain is of a pricking nature, as if thousands of splinters were imbedded in the tissues. Suffering increases with temperature extremes such as exposure to cold drafts of air or to hot weather. Clinical uses include Hairy Hoof warts, rectal fistulas, overuse of antibiotics, Actinomycosis, and other painful bleeding warts.

Nux vomica (Poison Nut)

Digestive upset from overeating rich food; constipation, colic or bloat; straining is a predominant concurrent symptom. Patient is hard working and intense; toxemia from conventional drugs, poisons, vaccines and spoiled food; acetonemia and ketosis; deranged digestion with disorders of liver or abomasum. Patient is better from warmth, rest and passing manure; worse after eating, mornings or becoming chilled.

Ovarian (Oophorinum)

Made from liquid expressed from the ovary of a healthy sheep or cow, clinically this medicine appears to enable the cow in proestrus to be ready to come back into estrus and increases tone of the reproductive tract. By palpation, the animal will be found to have a follicle on one ovary. Often there is a history of lowered conception rates due to delayed ovulation. Generally given in the 30X dose.

Palladium

Indicated in ovary difficulties, especially when right ovary is painful with relief when pressure is applied. The metal palladium has affinity for the right ovary attended with localized peritonitis within the pelvic organs. It is likely to benefit those cows and heifers showing irregular estrus. By palpation the animal will have follicular activity on the right ovary. They resent direct pressure on the right ovary or fallopian tube, but may enjoy a gentle massage of these same organs. They are better from pressure and rubbing (massage) and worse from exertion.

Phellandrium

The Water Dropwort plant has affinity for nerves, the chest and the udder. The pain in the teats is made better from nursing and milking. Right-sided bronchitis and mastitis call for *Phellandrium*. First nursing causes pain, but worse pain is experienced at cessation of nursing. In cows, hourly stripping out of the affected gland in conjunction with this medicine for right-sided mastitis may affect a cure.

Phosphoric acid

Acidum phosphoricum is characterized by pain and debility. Diarrhea leading to debility in young calves is a keynote. Fluid loss leads to weakness and in lactating cows there is sluggishness and a drop in milk production. Many symptoms are like *Phosphorous* only more extreme. All symptoms are worse from sour feed and improperly cured silage. In general, the patients are better keeping warm.

Phosphorus

This crude mineral has affinity for every tissue and organ of the body. Keynote symptoms are: thirst for large volumes of cold water; desire for moist juicy foods; restless, fearful, especially of loud noises and thunder storms, easily excitable; blood tinged mucous and blood in stool; bronchitis, dry cough and ketosis with loose manure; mastitis with fever and thirst; any illness from eating too much salt. Symptoms are often similar to *Bryonia* except the patient is more active. Cattle needing *Phosphorus* will become febrile during or shortly after calving. Other predominant symptoms are excessive bleeding after surgery, nosebleeds, anxiety when alone; relaxed rectum and anus after stool (cannot hold a thermometer), and great weakness after each stool. Complementary medicines are *Lycopodium*, probiotics, and B-complex vitamins. Reducing the amount of grain in the ration to one or two pounds of oats twice daily and offering all the clean mixed hay that she can eat will do much to settle the digestive tract.

Phytolacca decandra

Poke Berry plant has a profound affect on the glands, the breast, the bones and connective tissue. A hard, painful udder with moderate fever is keynote; the pain experienced radiates from the breast throughout the entire body; no wonder the cow kicks when we attempt to milk her. Swelling of the parotid glands and tonsils makes swallowing difficult; there is dark red discoloration to the throat. When mastitis diagnosis results from laboratory confirmation, *Phytolacca* combined with the appropriate nosode is effective in a high percentage of cases. Patients are worse form wet or cold weather and better from warm, dry weather.

Platanus occidentalis

Plantanus occidentalis is called the "lancet of the eye". It is used for cysts and lumps in the eyelids, especially Calazia and styes. It is usually used in low potency, two or three times daily. It can be administered locally as an eye wash of equal parts *Calendula* and *Platanus* tinctures, diluted one drop of each in twenty drops water. It bears its name well. (Entry by David Wember, MD.)

Platina

The metal Platinum is preeminently a female remedy. Strong tendency to paralysis, anesthesia, localized numbness and coldness are shown. In cattle, there is left ovarian inflammation with strong signs of estrus lasting for two or three days. Often a cyst develops on the left ovary. Breeding at proper times does not result in conception. Low potencies are useful in proestrus and higher potencies given at mid-cycle will help to regulate the entire estrus cycle.

Plumbum metallicum

The homeopathic preparation of lead triturated in milk sugar has affinity for any organ or tissue of the body that may be damaged from heavy metal poisoning. Keynotes are progressive muscle atrophy, jaundice, arteriosclerosis, and anemia. Extensor nerve (radial or tibial) and muscle paralysis of hind legs accompanied by constipation call for its use. An additional indication is heavy metal poisoning involving both the liver and kidneys. Paralyzed limbs are better from rubbing and stretching out.

Podophyllum peltatum

The characteristics of the *Podophyllum* (May Apple) state are diarrhea immediately after eating; thirst toward evening, and during fever; colic with retraction of the abdominal muscles generally beginning at daybreak; prolapse of the bowel and loss of strength after evacuation; morning diarrhea of dark, yellow mucus, with natural stool in the evening. The coat is soft, flabby with moisture and, in general, the symptoms are worse in the morning and better in the evening. It is an extremely useful medicine when stools contain large volumes of water. The term "pipe stream" diarrhea with little straining best describes the general symptoms. Early morning episodes of diarrhea during teething is a particular symptom. *Podophyllum peltatum* is most useful in the warm summer months and often brings rapid relief in young, growing animals.

Psorinum

See Nosodes

PTT (Posterior Pituitary, Thymus and Thyroid)

Mother tincture is made from bovine tissues of organic cattle. Has proven clinically useful in regulating thyroid disorders. It is postulated that postpartum livestock that were not cycling required endocrine stimulation. In conventional practice, *PTT* has been used to increase uterine contraction and promote milk let down. Homeopathic preparation should have positive effect on endocrine system. Most appropriate use is in animals that calved 30 days or longer and are not cycling.

Pulsatilla nigricans (Wind Flower)

In cattle, the animal looks healthy, is friendly but timid and prefers being with the herd; all body discharges are thick, creamy and yellow/green; mastitis is especially common at calving; postpartum metritis with creamy discharge; infertility and anestrus. Often there is a history of dystocia where the calf was not in the proper position for birth. All symptoms are improved from being outdoors and from fresh air. Symptoms may be worse just before dawn and just before sundown. *Pulsatilla* has been found useful during the period of parturition in animals; also for abortion; and with *Belladonna* in inflammation of the udder. *Pulsatilla* is also indicated when there is sinusitis with bland eye discharge.

Pyrogenium (Pyrogen)

The animal experiences a temporary increase in mastitis or other septic illness. Sick animals have either a rapid pulse with low temperature or a slow feeble pulse with a high temperature. Discharges, including mastitic milk, are dark and offensive. There may be straining as seen in cases needing *Nux vomica* and *Mercurius corrosivus*. History indicates much antibiotic use in the past, including dry cow mastitis tubes. Postoperative incisions become easily infected. In cattle, dose frequently after a miscarriage to help pass the afterbirth. Patients are worse from cold, damp weather and better from warm drinks and moving about.

Rescue Remedy

The combination of live Bach Flower Remedies is the medicine first indicated in accidents (like *Arnica*), and in shock (like *Aconite*). We recommend mixing three drops into four or eight ounces of water and dosing every 15 minutes until the animal regains consciousness. Rescue Remedy will calm cattle before surgery, internal exams and routine blood testing.

Rhododendron

The yellow Snow Rose from Siberia has symptoms that are always worse before a storm. Rheumatic joint pain especially in breeding animals calls for this prescription. Stifle injuries (gonitis) in bulls from over-work will respond to either *Ruta graveolens* or *Rhododendron*. Testicular swelling (orchitis) usually responds favorably to *Pulsatilla, Spongia tosta,* or *Rhododendron*. These animals are worse before a storm and better after the storm breaks.

Rhus toxicodendron

There is stiffness of muscles and tendons where pain is worse on beginning to move. Arthritic pain that is worse in cold weather; mastitis that is worse in cold, damp weather; muscle cramps from overwork or lying on cold, damp concrete; stiffness of tendons in lower legs after laminitis: all are common symptoms. All pains are better from warmth, stretching, and continuous motion. Useful in lactating does and ewes with sudden unexplained drop in milk production and in older breeding animals of all species with pelvic pain.

Ruta graveolens

Bruised bones and joints; pain in limb worse from touch; back pain better lying on it. The pains are worse at night and medicine is useful when animals become constipated after accidents and trauma.

In the bovine, we dose with *Ruta* for falls on wet concrete or in the gutter and after calving if there has been bruising to pelvic bones. Older breeding animals have relief from arthritic pain when *Ruta* is given. RRA (*Ruta, Rhus, Arnica*) ointment may be massaged into affected joints. In all species, it is helpful after bone injuries and with arthritic conditions that are worse in cold damp weather. Like *Rhus tox*, these patients are also worse from first motion, and better from continuous motion.

Sabina (Savine)

Made from the young fresh branch tops of the *Juniperus sabina* this remedy is indicated in cases where there is miscarriage with hemorrhage of bright red blood that remains fluid; dark, bloody discharge after calving; retained placenta; lack of uterine tone; and inflammation of ovaries. The pains in animals needing *Sabina* are from the sacrum to the pubis, somewhat like *Cimicifuga racemosa*. Skin symptoms of warts like *Nitric acid* and *Thuja* are common. These cattle are better in cool, fresh air than in warm stuffy barns.

Santoninum (Santonin)

Santoninum comes from the unexpanded flower of *Artemisia maritima* and homeopathic *Cina* is made from the same plant. In livestock, it has been combined with *Chenopodium* and *Abrotanum* to help in the control of roundworms, threadworms and strongyles. The three medicine combination may be dosed in 3X or 3C TID six days or continuously in the drinking water for six to 14 days.

Secale cornutum (Ergot of Rye)

Animal will be thin, scrawny, feeble, cachectic (chronically ill, emaciated) in appearance. Everything seems loose and open; no action, vessels flabby and all tissues lack healthy elasticity. Indicated when there are dark, watery postpartum hemorrhages. It has been curative in Navicular disorders when the feet are cool and the arterial pulses are weakened.

Rye grasses are apt to be affected with the Ergot disease in damp seasons, and when grown on damp, ill-drained lands. If breeding cows are turned onto pastures where infected grasses grow, they are liable to drop their calves prematurely. The opposite condition also can occur when calves are born on time, but the udder does not fill with milk so that the calves must be hand reared or they will starve to death. *Secale cornutum* will antidote some of the toxic effects of fescue pasture and fescue hay. Dose with 12X, 30X or 30C TID for seven days. Ergot has been known as a hastener of parturition.

Selenium

This trace element is necessary for healthy livestock. In much of the United States, soils are deficient and in other areas they may be an excess in the soil. Organs of the body most under its influence are genitourinary, musculoskeletal, hepatic, skin, hair and hooves. Dry cows benefit from once weekly dosing to prepare for calving. Older bulls with poor fertility or prostatitis will benefit from daily dosing. Any animal with weak bladder tone and the leaking of urine is a candidate for *Selenium, Causticum,* or *Sepia.*

Sepia (Inky Juice of Cuttlefish)

Impaired digestion, liver function, and decrease in blood supply to uterus and ovaries. Silent heat or infertility with tendency for organs to prolapse; tendencies for miscarriage, stretched ligaments and sagging reproductive tract. Patient has loss of natural affection; may not care for her offspring, and may lack interest in natural mating. Frequently observed is weakness of the bladder and hind limbs. *Sepia* will help tone the entire reproductive tract in preparation for estrus, also stimulating to liver and thyroid gland. Giving *Sepia* on day 21 postpartum results in improved conception rates and shorter calving intervals.

Silicea (Pure Flint)

Its action is deep and slow and will prevent the formation of scars. *Silicea* tends to soften scar tissue, especially fibrous tissue of reproductive tract (ovaries, fallopian tubes, uterus or cervix), and improves the integrity of connective tissue (ligaments, skin, hoof and hair). In cases of the reproductive tract there are often concomitant symptoms such as thin, hidebound condition with poor weight gain in spite of a high quality diet. Defects in absorption, digestion and assimilation of nutrients may also be observed. Injuries tend to suppurate for long periods of time. *Silicea* will help the body expel tiny bits of foreign material and promote hoof growth.

Spongia tosta

Roasted sponge has affinity for the thyroid, the heart, the throat, the testes and the lymph glands. It is useful in young calves with cough and swollen cervical glands where the cough is hoarse, suffocative and raspy. Adult cattle with swollen thyroid glands and erratic cycles will benefit when there is not a favorable response to *Sepia*. Breeding bulls with orchitis or painful epididymitis respond to *Spongia* or *Pulsatilla*. The cough of *Spongia* is "like a dull saw cutting a pine board." The patients are better lying horizontally and worse from cold, dry winds.

Staphysagria

The 3X homeopathic dilution can be added to the drinking water or applied as a spray or wipe on in cases of insect bites, a one percent solution is safe for people, livestock and pets. Higher potencies of *Staphysagria* are prescribed for urinary tract disease and postoperative incisions when the parts of the body have been stretched. These patients are very sensitive to pain and become angry at their surgeon, especially when the incision hurts and does not heal well. Like *Platanus* and *Symphytum* in low potencies, *Staphysagria* (in several potencies) can be useful to treat Chalazae of the eyelids. These patients are worse from the least touch of the affected parts, better from a warm rest.

Sulphur

Every animal alive will respond to *Sulphur* in one way or another. This is true for people as well. Homeopaths prescribe it when the well indicated medicine fails to act. *Sulphur's* diarrhea is often worse in the morning. Concurrent skin symptoms like ringworm, snow scold or lice may be present. All discharges are offensive like *Arsenicum album, Carbo vegetabilis,* and *Mercurius corrosivus.* However, the patient is hot rather then chilly and prefers cool places. He likes cold drinks (like *Phosphorus*) and experiences painless diarrhea (like *Podophyllum*). *Sulphur* has been most useful when allopathic or conventional drugs have been previously given to the patient with no success. In cattle with recurrent mastitis, medicines to consider are *Sulphur* and/or *Lac caninum* along with the appropriate nosode.

Symphytum officinalis

Pricking pains in bone; fractures—after *Arnica montana* has been given for bleeding; pain in eye after blunt trauma. In all species, *Symphytum* is helpful for ligament tears and for shin splints (periostitis). It follows *Ruta* well in ligament injuries. Because of the medicinal properties of *Symphytum* as it affects bones, fractures often respond quickly. Make very sure that proper alignment of bone fragments is accomplished before dosing with *Symphytum officinalis.* I have seen a few patients with crooked legs because casts were improperly applied. In cases where there is a doubt, give *Calcarea phosphoricum* until proper alignment is verified. Also choose *Symphytum* in cases where the cornea has a bruised appearance or there has been blunt trauma to the eye.

Thiosinaminum—Rhodallin

Thios has long been prescribed for scar tissue throughout the body, especially in the eye, the ear, the joints, the rectum, the uterus and the teat end. Dairy cows with teat injures from milking machines (or hooves) respond to this medicine both locally and orally. First aid medicines in teat injury are *Arnica, Hypericum,* and *Calendula. Thios* fits the convalescent stage where it softens granulation tissue and prevents the formation of scars.

Thuja occidentalis (Arbor Vitae)

This medicine, long known as a clearing remedy, has been found useful in the following conditions: cauliflower-like warts that bleed easily; hoof problems; ill effects from vaccination; ovaritis, especially on left side; and chronic illness. Animals are usually worse during cold, damp weather. Often there is a history of adverse reaction to vaccinations given during the last ten years.

Urtica urens

The Stinging Nettle has many homeopathic and herbal uses. A brief list of conditions calling for its use is: agalactia, bee sting, burns, gout, hives, kidney stones, rheumatism, spleen disorders, sickness from eating shellfish, sore throat, urticaria and worms. There are many more indications, but most often in livestock it is prescribed for udder edema and to improve milk letdown. Often heifers and young cows will hold up their milk until their calf nurses. When the milking machine is applied to the teats, there is little or no milk released. The two medicines to consider in such cases are *Ignatia*, and *Urtica urens*. Low potencies assist in drying up cows at the end of lactation.

Ustilago madis (Corn Smut)

Indicated for flabby, fluid-filled uterus after parturition; dark, brownish-black stringy vaginal mucus with poor involution of the uterus. May have pain in left ovary; worse with pressure and accompanied by light-colored diarrhea. There is often loss of hair and weakness or destruction of hooves.

Veratrum album

Young animals with prostration following copious diarrhea call for this prescription. Dose hourly in acute conditions. In sub-acute cases, dose four times daily for three days. Colicky animals with diarrhea often respond to *Veratrum album* or the mineral form *Arsenicum album*. Frequent dosing is required. It is important to examine the animal for areas of heat and coldness. Often, the chill is most obvious on the muzzle or the legs or feet.

Vespa crabro

The venom of the European hornet has similarities in affect with *Apis*. There are stinging pains in the left ovary and the right ear and eye. *Apis* has affinity for the right ovary. Puffy swellings of the face, neck and throat call for the use of *Vespa*. Skin affections are better from cold vinegar water baths.

Nosodes for Treatment & Prevention

The definition of nosode is: a product of disease obtained from any affected part of the system in case of illness and thereafter potentized. Animal Nosodes are homeopathic medicines prepared from the discharges of past patients. The response of the tissue to invasion by bacteria or viruses results in the formation of substances which are, in effect, the basis of the nosode. Mucus (or sputum) from the throat, or pharynx, or nasal passages, mixed with distilled water and grain alcohol, will yield a mother tincture. A homeopathically trained pharmacist or veterinarian will prepare in step-wise fashion from the mother tincture a respiratory disease nosode in 30C potency. The 30C medication may be prescribed for either prevention or treatment, along with other carefully prescribed homeopathic medications.

We prepare milk Nosodes from individual farms by culturing milk samples. If we are making a nosode for use in the entire herd, we always request a bulk tank milk sample. The bulk milk tank sometimes yields different species of bacteria than those recovered from individual quarter samples. Milk samples from the highest SCC cows often yield the best bacteria growth on culture plates. Milk Nosodes in 30C or higher potencies are generally very effective for animals on the farm of origin. The term "isode" may be used for a sample taken from the patient and given back to the same patient. Isodes and Nosodes are always prepared according to HPUS (Homeopathic Pharmacopoeia of the United States).

All Nosodes should be prescribed by a licensed veterinary practitioner who has been thoroughly educated in homeopathic medicine. They are useful in association with single homeopathic prescriptions.

Bacillinum Nosode

Used for yearlings and two-year-old cattle with signs of ring-worm or other dermatitis (lice, mange). Associated medicines may be *Natrum mur, Sepia* or *Sulphur.*

Bovine Calf Pneumonia Nosode

This nosode is prepared from respiratory secretions produced by calves with this condition. Useful in prevention, the general dosing in one- or two-week-old calves is twice daily for three days, then once weekly for three weeks. Respiratory infections are common in housed young animals. Following *Aconite* (for three days) with bovine calf pneumonia nosode will often prevent pathogenic tissue changes in the lungs.

Bovine Foot Rot Nosode and Hairy Hoof Wart Nosode

Bovine Foot Rot nosode and Hair Hoof Wart nosode from fluids, tissues and cultures are useful in the prevention and control of hoof diseases in mature cattle.

Bovine Virus Nosode

Contains several common viruses (IBR, BRSV, P13, plus two strains of BVD) with *Haemophilus som* and aids in the prevention of these diseases in closed herds and organic herds. Associated medicines for treatment are *Phosphorus, Echinacea, Arsenicum album,* or *Thuja occidentalis.*

BRSV (Bovine Respiratory Syncitial Virus) Nosode

BRSV nosode, like Bovine Calf Pneumonia nosode, is prepared from respiratory secretions of clinical cases. The medicines will aid in the prevention of respiratory disease in young cattle that must be in confinement housing. Bovines are most susceptible from two months of age to two years of age.

E. coli or Colibacillinum Nosode

Made from the culture of bacteria in white calf scours. Used when there is a history of calves breaking with diarrhea at two to six weeks of age. *E. coli* nosode given shortly after birth will provide some energetic resistance to disease. Young calves that are crowded or under stress from other environmental influences become susceptible to Gram-negative bacteria. This nosode is not a panacea, but will reduce the incidence of enteritis and improve survival rates in herds.

E. coli Mastitis Nosode

This nosode is made from mastitis cultures containing gram-negative organisms. Dosing herds with this medicine along with the simillimum will aid in the prevention and control of Coliform mastitis and associated diseases.

Mixed Mastitis Nosode

Mixed Mastitis nosode contains cultures of both gram-positive and gram-negative organisms that have been associated with mastitis in dairy cattle. This medicine has been prescribed in herds with elevated somatic cell counts where herd milk cultures have grown a variety of gram-positive and gram-negative organisms.

NFE (New Forest Eye) Nosode

This nosode is made from the secretions and tears of actual cases of Bovine Pink Eye disease. For prevention and control of NFE-like infections in season, dose daily for five days and then once weekly for five weeks. Out of season dose daily for five days and then once monthly for three months.

Staph Aureus Mastitis Nosode

Originating from a milk culture, it has been used in thousands of cattle as an aid in the prevention and control of mastitis associated with this organism. This nosode, combined with other carefully prescribed homeopathic medications, may result in a clearing of pathogenic bacteria from the mammary gland.

Streptococcus Mastitis and Streptococcus Agalactia Mastitis Nosodes

These two Nosodes are prepared from the cultures and prescribed along with the simillimum for control and prevention of mastitis. Some cases of Streptococcal mastitis respond to *Phytolacca decandra* dosed TID and the appropriate nosode BID.

Sycotic Co. Nosode

Sycotic Co. nosode, also called Sycotic Bacillus, is indicated for irritation of mucous membranes. Some indications are food allergies, chronic bronchitis, chronic catarrhal enteritis, conjunctivities, Coccidiosis, Johnes' Disease, and warts in and around the rectum. This is one of the Bowel Nosodes which were developed in England by Drs. Bach and Patterson. Associated medicines are *China officinalis, Hydrastis, Mercurius corrosivus,* and *Nitric acid.*

Tuberculinum Aviare Nosode

This is a useful remedy for some herds where chronic bronchitis and pneumonia have been a problem in calves who are not responding to the simillimum.

Vermont Teat Scab Nosode

Manufactured from actual teat tissue containing a firm nodular eruption. This medicine has been used along with Bovine Herpes Mammilitis nosode to help remove painful lesions and skin tags from the bovine teat and udder.

9.

Beginnings

Changing to a Holistic Practice

To make an ending is to make a beginning. In 1981 my dissatisfaction with conventional veterinary practice had reached its zenith. When I first heard about homeopathy, the name was completely foreign—let alone the definition of the term. A new door opened with an opportunity to sit for two weeks under the teaching, and in the presence, of George Macleod, MRCVS. By the end of that course, I learned the definition and a whole lot more.

My view of conventional veterinary practice at that time was falling short of its advertised success. I had been taught in veterinary school that doctors could cure disease. Our efforts and our medicines were to make animals better and keep them healthy. Except for the surgical triumphs that I was experiencing, I felt that the rest of my veterinary practice fell short of the ideal.

I refused to "get another hobby" or "to buy a second home at the shore" to get away from my practice. Besides seeing this as taking the easy way out of frustration, my bank account couldn't survive an expensive hobby or a second home. Instead, I threw myself into learning a new system of medicine, one that I was not taught in veterinary school.

Holistic Steps

Homeopathy is at the core of holistic medicine because it endeavors to address the whole patient. Whether we spell it *holistic* or *wholistic*, to me the meaning is the same. Stepping back far enough to view the animal in his environment is what all holistic practitioners attempt to do. Taking the time to record the totality of symptoms forces one to see the larger picture.

One of my first large animals cases was an American Saddlebred gelding named "Smokey." This young (three years old) driving horse had just been given a death sentence. The diagnosis was Navicular disease. The conventionally-trained veterinarian presented three treatment options, none of which the owners wanted to consider: cut the nerves to the front feet so that the animal will not feel the pain, sell the horse without a cure, or destroy the animal.

The person on the phone was Amish farmer and horse trainer, Roy Miller. Roy had sold the three-year-old gelding to his uncle, Sam Miller, who was also an Amish farmer. Both Miller families really liked "Smokey" and did not want to lose him. They also agreed that a neurectomy (cutting the nerves) on such a young horse did not seem fair to the animal. "Smokey" was otherwise a healthy horse and, if recovered from the lameness, could gallop around the pasture for 20 or more years. Nerved horses no longer feel anything in their feet; after a total neurectomy they can sustain serious injuries to their hooves and not know it. The Millers did not allow the surgery. Instead, they chose homeopathy.

"Smokey" responded well to the homeopathic prescription of *Rhus tox;* he became sound within two weeks. The follow-up prescription of *Secale cornutum* appeared to be keeping him sound. After several weeks of rest, the Millers tried light driving and the horse seemed to enjoy it. Very gradually, the gelding was brought back to full work. Within a short period of time the Millers announced that "Smokey" was cured. After I examined the animal, I agreed with their evaluation. Before long I was being asked if the 'little white pills' that "Smokey" was taking might be used on cows.

Farmers with Notebooks

Roy asked if I could be willing to perform a herd health check on his Holstein dairy cows. A herd check consists of internal and external examinations for health and fertility. Roy's experience with "Smokey" was changing his thinking about how to properly care for farm animals. It seemed that he was ready to get off the antibiotic-steroid-hormone wagon and try something different.

One day as I was slipping into boots and coveralls in the milk house, I noticed farmers coming down the lane in vans and carriages. This entourage, armed with pencils and notebooks, were led by none other than Sam Miller. These Amish farmers began following me from cow to cow, writing down everything that I said. Roy Miller later explained that the farmers with the notebooks were his neighbors and his uncle's neighbors, all of whom wanted to witness a homeopathic veterinarian at work. Before long the homeopathic veterinarian was working on many farms in Lancaster County and the surrounding countryside. The medicines that worked for horses also worked for mules and cattle. Later on they witnessed it working for their sheep, goats, llamas, chickens, dogs and cats. Soon they were asking my wife and me to make up kits of homeopathic medicines for the barn. Many wanted to know which medicines could be taken by their families and themselves.

Saying Goodbye to Conventional Veterinary Practice

That was the beginning of the end of my conventional veterinary practice. Neither I nor the farmers knew anything about organic farming at that time. Thanks to *Acres U.S.A.* we soon heard the terms *organic* and *sustainable*. For a number of those first Amish and Mennonite farm families, changing to a more natural system for the whole farm became the logical next step. Within a few years, we heard about grazing systems on organic farms and those were adopted as well. Having made the transition to a natural system of health care for their livestock, the farmers were able to make other progressive changes more easily.

In 1990, the Clark Veterinary Clinic, Inc. was servicing two certified organic dairy farms, and by 1995 there were over forty farms in transition, on their way to being certified organic—and some of those had already made it. Nearly half of those forty farms were also in transition to a grass-based system of farming. Today the number of organic grass-based farmers that call our office is in the hundreds.

Farmers are accustomed to veterinarians who are in a hurry. When the Lancaster County Amish heard that there was a practicing holistic veterinarian who charged by the hour, the concept, at first, was hard to accept. The conventional vets charged for medicines and procedures. Many farmers looked at it like Daniel J. Stoltzfus, who thought that it might be OK to have a holistic vet charging by the hour if he moved as the "fire engine vets" did. Daniel would say, "Bacteria do not just grow, they multiply; one become two, two become four, four become eight, and so on. Fire engine vets must hurry, because the longer they wait, the more antibiotics they must administer. If they don't give drugs quickly, the bacteria will outnumber the antibiotics."

In Daniel's own words, "Most veterinarians come in the drive very fast, give the cow a shot and leave as quickly as they arrived. I had never had one on the farm who was not in a hurry. Sheaffer drove in the lane slowly; he was looking around. When he got out of the vet truck he said hello and talked about the weather. I thought, I'm paying him $90.00 an hour to look at my farm and talk about the weather?

"It was the same with the cows. He didn't reach inside and proclaim pregnant or open, but he looked the cow all over. He told me that this cow has had a stomach disorder and the next cow was under kidney stress, and the third cow was having trouble with the liver. He told me more about each cow on that first herd health exam than I ever thought anyone could find from an internal and physical exam. He mentioned that the whole herd had indications of possible mineral imbalance."

Daniel helped himself and his veterinarian by having all the cows lined up in stalls and having already prepared information on each cow. I really appreciated this effort, and we worked together to make that our standard. It resulted in a more profitable farm visit for all concerned. After spending two hours battling traffic, I was happy to find the cattle lined up and the farmer ready to go.

Daniel and his neighbors soon became regulars when it came to using homoeopathic medicine. Every four weeks found me on his farm for two or more hours observing and examining cows, calves, mules, horses, chickens, goats and pigs. Together we studied health records, soil test reports, and forage analysis reports and we talked about how we could use homeopathic medicine to its best advantage for all the animals on his farm. Often, other subjects related to alternative and complementary medicine were discussed and explored.

Dispersion

Lancaster County, Pennsylvania is often called the garden spot of the world. In truth, soil fertility has been very good for the past 300 years. Farmers from southern Germany and other regions of central Europe immigrated to the United States in the eighteenth century. Those Anabaptists with Amish and Mennonite heritage settled in regions of Pennsylvania, Indiana and Ohio. The two largest settlements were Holmes County, Ohio and Lancaster County, Pennsylvania.

By the end of the twentieth century every acre of Lancaster County was being used and every farm had been subdivided again and again until the land could not support additional farms. The exodus to New York, Indiana, Wisconsin and other parts of Pennsylvania by the Anabaptist farmers was occurring at a rapid rate by the late 1990s. To accommodate this migration, the circumference of my practice area increased along with the new locations. I decided to limit my southern borders at the Mason-Dixon line, but took on new farms in the north and west. When I thought that a farmer was serious about a transition to sustainable farming systems, I visited his new location. From 1993 through 1999 I was driving over 35,000 miles a year servicing sustainable and organic farms.

Boldness to Speak Up

Wonderful results are experienced as the result of homeopathic medicine. Some people may not see the connection unless they reflect on it or God reveals the fact to them. Anabaptist farmers are usually shy about telling others what they are doing. I had been doing work for Jacob for about eight years. His herd health had improved, his milk production was up, and the entire farm was now in pasture and hay. As I drove over the hill to the farm buildings, I had a beautiful view. His success was very apparent to me. I questioned him, "Do any of your brothers or sisters ask about the different way that you are farming? Do they ask what kind of medicine you are using on your livestock?"

His answer was a simple, "No." Implied in that answer was, "Unless I am asked, I am not going to volunteer any information." Fortunately, Jacob's son, John, watched his dad and learned from his actions and not his words. Years later, John has an equally beautiful grass-based organic dairy farm in another part of the state.

On another occasion I asked a farmer if he would be willing to talk to other farmers in a meeting about his experience using homeopathic medicines. He replied his shyness prevented him from speaking before large audiences. He indicated that perhaps he would have the courage to speak one-on-one with a farmer, but he could never stand up before a group. Two farm accidents and many homeopathic doses later, he was a different person. Although he never volunteered to speak at a homeopathic meeting, he did speak at several large meetings of farmers. He now had the boldness to speak. When a person takes homeopathic medicines and experiences the results, something intangible occurs to his personality.

Thank You Doctor for the Nice Medicine

While performing a herd check for her herd of Jerseys and Holsteins, Mrs. H. mentioned that four of her six children were home from school with pink eye. "The eye drops we were given by the pediatrician aren't helping, and the children hate them. The two oldest girls must lie on the smallest children to hold them down so that we can put the drops in and afterwards they scream. It must really hurt."

I stopped examining cows for the moment and returned to the truck. After a search I found one bottle of homoeopathic eye drops containing *Euphrasia, Belladonna,* and *Mercuris.* The medication had a human label, and I often prescribed it for animals. Handing the bottle to the oldest daughter, I said, "Try this."

One hour later I had finished examining the cows. After scrubbing my boots and washing my hands, I looked up to the yard in front of the house. There stood four children in a straight row facing me. Like children reciting a Sunday school recitation, in unison they said, "Thank you doctor for the nice medicine."

Homeopathic eye drops did not cause pain, and the children were well in a short time. A holistic veterinary practice helps every living being on the farm.

Glossary

Acute Disease. Illness of rapid onset.

Allopathy. Conventional treatment of diseases with medicines unrelated to the disease and which do not target the underlying problem. For instance, using an antibiotic for an infection.

Centesimal. Denoted as 'C'. Scale of dilution of a remedy indicating one in one hundred.

Chronic Disease. Illness that is well established and of long standing that has not been remedied.

Colostrum. First milk produced after birthing, rich in antibodies and nutrients.

Concomitant Symptom. A symptom secondary to the illness presenting itself. Used to aid in prescribing homeopathic remedy.

Constitutional Remedy. A remedy that considers the patient's entire make-up and not just the illness presenting itself. It affects all parts of body including the mind.

Conventional Medicine. What is currently used in veterinary practice, antibiotics, drugs, vaccines, surgical procedures, etc. Focus on illness or problem, not on whole animal.

Cyanosis. Tissues and membranes turn blue.

Decimal. Denoted as 'D'. Scale of dilution of a remedy indicating one in ten.

Dysentery. Inflammation of the intestines with passage of mucus, bowel lining or blood with feces.

Holistic. Regarding the whole body, spirit, mind in the context of the environment. Sometimes called 'wholistic'.

Homeopathy. A system of treating disease with remedies that produce, in a healthy animal, symptoms like those of the patient.

Materia Medica. A book that lists homeopathic medicines, their properties and the symptoms for which they should be used.

Miasm. A term devised by Dr. Hahnemann to describe susceptibility to infective agents.

Modality. Influence by temperature, weather, time of day, movement, etc., on symptoms.

Mother Tincture. Undiluted medicine made from the source material to form a remedy. Generally in alcohol solution. Remedies are diluted from this source.

Nosode. Remedies created from the infected tissues, discharges or organisms of a disease.

Palliative. A medical treatment whose aim is to reduce symptoms.

Parturition. Giving birth.

Polychrest. A constitutional remedy, one that has affect on entire organism.

Potency. The dynamic action of a remedy that has been diluted and succussed, denoted by C, D, X with a number indicating the stages it has passed through.

Prognosis. Best guess as to outcome of disease or illness.

Repertory. A dictionary of symptoms that includes the homeopathic remedies or medicines for particular symptoms.

Septicemia. Pathogens in the blood stream.

Side Effects. Unwanted reactions to drugs or medications.

Succussion. Shaking of dilutions of homeopathic remedies in order to potentize them. Done under controlled conditions.

Symptom. The visible ways in which a patient is fighting illness or disease. What the homeopath uses to decide which medicine/s are appropriate.

Symptom Picture. All of the symptoms and signs that a patient is showing in reaction to disease or illness, used to decide how to proceed homeopathically.

Tincture. Solution or alcohol or water and alcohol, usually dropped in mouth or water for patient.

Topical. Medicine or remedy applied to the body surface.

Trauma: A damaging injury.

Trituration. Manufacturing process whereby insoluble material is ground with milk sugar before potentizing with a liquid. The level at which the transformation takes place varies from remedy to remedy. Usually it is somewhere between a 4X and a 12X potency.

References
& Resources

Books and Periodicals

Boericke, W. *Homeopathic Materia Medica with Repertory*. 9th Ed. Philadelphia: Boericke and Runyon, 1927.

Clarke, J. *Dictionary of Materia Medica, Vols. 1, 2, and 3*. Saffron-Walden, England: Health Sciences Press, 1900.

Cortese, Victor S. Lecture. "Proceedings of the Pennsylvania Veterinary Medical Association," 115th Annual Meeting. Lancaster Host Resort, Lancaster, PA. 1997.

Currim, Ahmed N., editor. *The Collected Works of Arthur Grimmer, M.D.* Norwalk, Connecticut. Hahnemann Inter-national Institute for Homeopathic Documentation, 1996.

Day, C. *The Homeopathic Treatment of Beef and Dairy Cattle*. Beaconsfield, England: Beaconsfield Publishing, Ltd., 1995.

Dossey, L., executive editor. "Alternative Therapies in Health and Medicine." Encinatas, California: Vols. 2-5, 1997-2001.

Hahnemann, S. *Materia Medica Pura, Vol. II. Ledum-Verbascum.* Hahnemann Publishing Society, England, 1881.

Hering, C. *The Guiding Symptoms of Our Materia Medica.* Philadelphia: The Estate of Constantine Hering, 1888.

Kent, J. T. *New Remedies; Clinical Cases, Lesser Writings, Aphorisms, and Precepts.* New Delhi: B. Jain Publishers, 1926.

Kent, J. T. *Repertory of the Homeopathic Materia Medica.* New Delhi: B. Jain Publishers, 1945.

Macleod, G. *The Treatment of Cattle by Homeopathy.* Saffron Walden, England: Health Science Press, 1981.

Macleod, G. *Pigs: The Homeopathic Approach to the Treatment and Prevention of Diseases.* Saffron-Walden, England: C.W. Daniel Co. Ltd., 1994.

Merck Veterinary Manual. 8th edition. Rahway, New Jersey: Merck and Company, Inc., 1998.

Moore, J. *Horses Ill and Well: Homeopathic Treatment of Diseases and Injuries.* London: James Epps and Co., 1873.

Murphy, Robin. *Homeopathic Medical Repertory.* Pagosa Spring, Colorado: Hahnemann Academic of North America, 1993.

Homeopathic Remedy Guide. Blacksburg, Virginia: Hahnemann Academy of North America, 2000.

O'Connor, J. *The American Homeopathic Pharmacopeia.* 4th Edition. Philadelphia: Boericke and Tafel, 1890.

Salatin, J. *Salad Bar Beef.* Swoope, Virginia: Polyface Farm Inc., 1995.

Studdert, Virgina P., and D. C. "Blood." *Baillieres Comprehensive Veterinary Dictionary.* Darien, Illinois: W. B. Saunders and Co., 1997.

Sullivan, Andrea. *A Path to Healing: A Guide to Wellness for Body, Mind and Soul.* New York: Main Street Books, Doubleday, Co., 1999.

Winston, Julian. *The Faces of Homeopathy.* New Zealand: Great Auk Publishing, 1999.

Winston, Julian. *The Heritage of Homeopathic Literature.* New Zealand: Great Auk Publishing, 2001.

Yasgur, Jay. *Yasgur's Hoemoepathic Dictionary and Holistic Health Reference*. 4th Edition. Greenville, PA: Van Hoy Publishers, 1998.

Contacts for Information on Homeopathy

American Holistic Veterinary Medical Association, 2214 Old Emmorton Rd., Bel Air, Maryland 21014, phone (410) 569-0795, fax (410) 569-2346, e-mail <AHVMA@ compu serve.com>.

National Center for Homeopathy, 801 North Fairfax St., Suite 306, Alexandria, Virginia 22314, (703) 548-7790, fax (703) 548-7792.

Homeopathic Medical Society of the State of Pennsylvania, Box 353, Palmyra, Pennsylvania 17078, fax (717) 838-0377.

Index

abdomen, distended, 146, 149; cramping, 149; pain around navel, 157

abomasal displacement, 107-108

abomasum, disorder of, 164

abortion, 168; prevention of, 103

Abrotanum keynotes, 140; for worms, 137

abscess, 81-82, 162

acetonemia, 164

acidophilus, for mastitis, 97

Aconite, prior to travel; to treat respiratory virus, 60; for fever, 70; for fear and fever, 77; for mastitis, 94-95; for respiratory affections, 131

Aconitum napellus, 8; preoperative use, 14; keynote symptoms, 24-25, 140-141; for dehorning, 115; for laminitis, 125; for navel ill, 127; for NFE, 128

Acres U.S.A., 30, 181

Acres U.S.A. Conference, 46

actinobacillosis, 158

actinomycosis, 129, 164

acute illness, 77

affection, loss of, 170

agalactia, 173

aggravation, 78

allopathy, 6

American Holistic Veterinary Medical Association, 30

American Institute of Homeopathy, 64

anemia, 151, 157, 166

anesthetics, 8

anestrus, 13, 130, 145, 149, 155, 167

animal protein, in feed, 45-46

antacids, 123

anti-convulsants, 8

anti-inflammatories, xiv, 8

anti-prostaglandins, 8

antibiotics, xiv, 8, 10; effects of, 58-59

Antimonium tartaricum, for respiratory affections, 131; keynotes, 141

antioxidants, 132

anxiety, 165

Apis cow, 43

Apis mellifica, 43; for mastitis, 95; for stings and bites, 108; for luteal cysts, 126; for urinary tract, 135; keynotes, 141

Arnica montana, 8; as beginning remedy, 30-33; for trauma and bruising, 70, 77; postoperatively, 107; for bruises, 110, 111; for calving paralysis, 104, 112; for dehorning, 115; for hardware disease, 121; for hoof fistulas, 123; for toothache, 129; for teat injuries, 132; for ulcers, 134; for wounds, 138, keynotes, 141

Arnica ointment, for watery mastitis, 96

Arrowroot Standard Direct, 126

Arsenicum album, for stings and bites, 108; for diarrhea, 115; keynotes, 142

arteriosclerosis, 166

arthritis, 108, 168; pain of, 160, 169

B-complex vitamins, 165; for mastitis, 96, 97; for ketosis, 108, 125; for stomatitis, 132

Bacillinum Nosode, 175

back, pain, 149

bacteria, on farm, 54-61; in milk, 55; control of, 99

bacterial infection, 151

Baptisia tinctoria, keynotes, 142

Belladonna, for mastitis, 81, 94; for heat exhaustion, 121-122; for tetany, 119; keynotes, 142-143

Bellis perennis, for bruising, 110, 111; keynotes, 143

benzimidazoles, 138

Berberis vulgaris, and immune system, 56; for urinary tract, 135; keynotes, 143-144

beta carotene, for stomatitis, 132

BGH (bovine growth hormone), 34

biliary disorders, 143

bio-industrial systems, x

biodiversity, xi

biosecurity, 59

bites, 108-109, 160

black cohosh, 149

Blackie, Margery G., 88

bladder, irritation, 146; pain in, 161; weakness, 170

bleeding, 141, 157

bloat, 53, 67, 70, 109, 149, 160, 164

bloating, 146

blood, 165; failure to clot, 150; poisoning, 153

blue cohosh, 147

Boericke, William, 29

Boericke's Materia Medica, 152

boils, 151

bone spavin, 144

bones, 145, pain in, 172

bovine foot rot, 110-111

Bovine Calf nosode, 28

Bovine Calf Pneumonia Nosode, 175

Bovine Foot Rot Nosode, 175

Bovine Virus Nosode, 175

bowel, prolapse, 167

brain, fluid on, 154

Bright's disease, 134

British Homeopathic Journal, 12, 13

bronchitis, 111, 164; right side, 165; chronic, 177

BRSV, 175

BRSV (Bovine Respiratory Syncitial Virus) Nosode, 175

bruised bones, 169

bruising, 111, 141, 143

Bryonia alba, and fever, 25; for constipation, 114; for hardware disease, 121; for respiratory affections, 131; keynotes, 144

BST, 136

burns, 146, 173

BVD, 35, 43, 51, 56, 57, 161, 175

Calcarea carbonica, and immune system, 56; for fertility, 70; for mastitis, 94, 95; for fractures, 117; for milk fever, 127; keynotes, 144

Calcarea fluorica, for arthritis, 130; keynotes, 144

Calcarea ostrearum, 117

Calcarea phosphorica, and fractures, 14, 118; and pneumonia, 17; for fertility, 70; for new calves, 104; for diarrhea, 116; for luteal cysts, 126; for ligament injury, 126; for milk fever 127; for stomatitis, 131; keynotes, 144-145

calcium, 119; deficiency, 70-71; for stomatitis, 132

calculi, 143

Calendula, as surgical treatment, 13; for trampled udder, 32; for scar tissue, 103; for Chalazion, 113; for dehorn-

kidneys, problems of, 143-144; poor function of, 160
labor, 147
Lac caninum, keynotes, 159
Lachesis muta, for coliform mastitis, 96; for gangrenous mastitis, 96; for stings and bites, 109; keynotes, 159
lactic acidosis, 163
Lacticum acidum, keynotes, 159
lameness, 145; shifting, 147
laminitis, 125-126, 141
laryngitis, 151
Lathyrus sativa, keynotes, 159
laxatives, 123
Ledum palustre, for injection site, 108; for eye injury, 117; for hoof fistulas, 123; for insect stings, 133; keynotes, 160
legs, swelling, 163
Leopard's Bane, 8
Leptospirosis, 35, 134
leukosis, 43, 57
lice, 132-133, 172
ligament injuries, 126; tears, 172
limbs, trembling, 161
liver, disorder of, 164; remedy, 148; inflammation of, 155; poor function of, 160; function, 170
luteal cysts, 126
Lycopodium clavatum, for ketosis, 70, 108, 125; for mastitis with ketosis, 95; for abomasal displacement, 107; keynotes, 160
Lyme disease, 133
lymph gland, hard, 157
lymphatic tissues, of throat and neck, 157
Macleod, George, xiii, xiv, 24, 28, 153, 179
Macy, Dennis W., 38
Magnesia carbonica, for milk fever; keynotes, 160
Magnesia phosphorica, 31; postsurgical use, 14; for tetany, 119;

for milk fever, 127; keynotes, 160-161
magnesium, 119
magnets, to treat hardware disease, 120-121
mammary glands, 159
marigold, 145
mastitis, 3, 12, 16, 24, 70, 81, 93-105, 127, 141, 143, 151, 154, 157, 159, 161, 165, 166, 167, 168; keynotes for acute case, 94; coliform, 95-96; gangrenous, 96; watery, 96; hard quarter, 141; septic, 96-97, 159; right side, 165
Mastoblast, 73, 100, 161; to lower SCC, 102
Mastoblast H.P., indications for use, 100
Mastocream, 96
Materia Medica, 4, 5, 29, 139-177
Materia Medica with Repertory, 29
membrane dryness, 144
Merck Veterinary Manual, The, 103
Mercurius corrosivus, for coccidia, 114; for diarrhea, 116; for stomatitis, 132; keynotes, 161
Mercurius solubulis, for stomatitis, 132; keynotes, 161
metabolism, depressed, 148; regulation of, 157
metritis, 145, 151, 167; septic, 159
Mezereum, keynotes, 161
miasm, 83-84
miasma, 23
milk, replacer, 44-45; viscous, 147; bloody, 157; inability to digest, 160, 163; letdown, 167, 173; drop in production, 169
milk fever, 127, 144, 145, 152, 160, 161
Milk of Magnesia, for bloat, 109
Milk Nosodes, 174
milking machines, and disease, 57
Miller, Roy, 180, 181

palliation, 2-3
paralysis, 131, 147; calving, 103-104
paralytic affections, 159
parasites, 131
Paronychia, 110-111
parotid gland, swelling, 166
pathogens, on farm, 54-61
Patient, Not the Cure, The, 88
pelvic inflammation, 103
pelvic inflammatory disease, 149
pelvic pain, 149, 169
pelvic weakness, and *Causticum,* 48
Penn State University, xiii, 139
pericardium, fluid on, 154
peritonitis, 164
Phellandrium, keynotes, 165
phobophobia, 150, 151
phosphate of soda, 163
Phosphoric acid, keynotes, 165
Phosphorus, 9; for ketosis, 108, 125; for dehorning, 115; for diarrhea, 116; for respiratory affections, 131; for ulcers, 134; for urinary tract, 135; for wounds, 138; keynotes, 165-166
Phytolacca decandra, 13, for acute mastitis, 70, 94; keynotes, 166
Phytolacca ointment, for watery mastitis, 96
pink eye, 131, 176, 184
placenta, 147, retained, 169
Platanus occidentalis, for Chalazion, 113; keynotes, 166
Platina, keynotes, 166
pleuritis, 3, 121, 150
Plumbum metallicum, keynotes, 166-167
pneumonia, 177; virus, 151
pododermatitis, 110-111
Podophyllum pelatum, for diarrhea, 115, 116-117; keynotes, 167
polycrest, 88
pomegranate, 152

poor digestion, 163
post-estrus, 71
postpartum, 171; hemorrhage, 170
prescribing, 75-91
probiotics, 27, 108, 151, 165; for calving paralysis, 112; for indigestion, 123; for ketosis, 125; for stomatitis, 132
proestrus, 70, 71, 164
prolapse, 170
prostaglandin, 8, 136, 162
proud flesh, 152
Pseudomonas, 12, 55
Psoric, 83-85
Psorinum, 167
PTT (Posterior Pituitary, Thymus and Thyroid), for luteal cysts, 126; keynotes, 167
Puerperal fever, 151
Pulsatilla nigricans, cow, 42; and immune system, 56; in proestrus, 71; for fertility problems, 95; to bring on heat, 102; to clear womb, 104; for retained placenta, 103, 112; keynotes, 167-168
pulse, rapid, 140, 143
pyrantel, 138
Pyrogen, for coliform mastitis, 96; for gangrenous mastitis, 96; for watery mastitis, 96; for septic mastitis, 97
Pyrogenium, for navel ill, 128; keynotes, 168
rectal fistulas, 164
rectal pain, 164
rectal sleeve, 43; and disease transmission, 57
rectum, relaxed, 165
Reece, Michael, 142, 154
renal disorders, 143
renal pain, 135
reproduction, problems of, 102-105
reproductive problems in cows, 42-43

uterus, 147; fluid in, 173
Uveitis, 124
vaccinations, 8, 35-40, 164; effects
    on herd, 26-28; ill effects of,
    173
vagal indigestion, 133
vaginal mucus, 173
Veratrum album, keynotes, 173
vermifuge, 152
Vermont Teat Scab Nosode, 177
vertigo, 152
Vespa crabro, keynotes, 174
veterinary homeopathy, x
veterinary practice, holistic, 182
Veterinary Materia Medica, 139
Veterinary Product News, 38
viral conditions, 161
viral infection, 151
viremia, persistent, xiv
viruses, on farm, 54-61; control
    of, 60-61
vitamin A, for stomatitis, 132
vitamin C, 27, 151; for mastitis,
    96, 97; for stomatitis, 132
vitamin D, for stomatitis, 132
vitamin E, and immune health, 36;
    for stomatitis, 132
Ward, James W., 15
warts, 136-137, 169, 173; rectal,
    177
Washington State University, 57
Washington State Veterinary
    College, 43
Water Hemlock, 148
watery mastitis, 96
weakness, 147, 148, 152, 157, 158
weaning, 156
Wember, David, 113, 166
West Nile Virus, prevention, 61
white hops, 144
white line disease, 125
Wilkinson, J. Garth, 153
Williamson, A. V., 89
Winston, Julian, 15, 140
Winter Dysentery, 116-117
worm seed, 149
worming, 158

worms, 137-138, 173
wounds, 138; open, 145; infected,
    153; puncture, 155, 160
Wynn, Susan, 47
Young, Arthur, 64

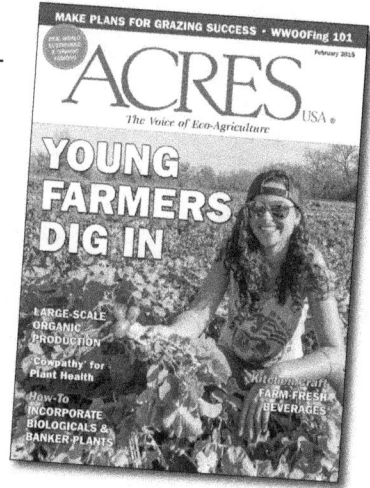